ANTARCTIC UPPER ATMOSPHERE

I0480016

R. PURUSHOTTAM BHAWRE

INDIA · SINGAPORE · MALAYSIA

Notion Press

No.8, 3rd Cross Street,
CIT Colony, Mylapore,
Chennai, Tamil Nadu – 600004

First Published by Notion Press 2020
Copyright © Dr. Purushottam Bhawre 2020
All Rights Reserved.

ISBN 978-1-64951-827-9

Dedicated to

Indian Base Station

"Maitri" Antarctica

&

RASD,

National Physical Laboratory (NPL)

New Delhi

CONTENTS

PREFACE

EARTH is part of solar system and it has some special features, which makes it beautiful and has conditions conducive for the humans to survive on it. It is well known that the main source of energy for the Earth is Sun. So any changes in, of and on the sun, affects the Earth's environment also called as Space weather. Gases and dust particles, the combination of which forms a gaseous envelope of Earth. Any changes in Earth's atmosphere may affect the Earth and in turn, the human life. Like this, there could be many phenomena which could spontaneously occur on the Earth's, in turn affecting the human life directly (seismic activity, variation in Earth's magnetic field, increase in temperature due to greenhouse effect).

The upper part of the earth's atmosphere is known as ionosphere and magnetosphere. The ionosphere consist the ionizing particles, which are generated by the solar photo ionization process. Basically the ionization is produced in the earth's atmosphere by a wide spectrum of solar X-ray and extreme ultraviolet (EUV) radiation. The distribution of ions in the ionosphere is not even, so according to distribution of ions, it is further divided into sub layers, known as D, E, F1 and F2. Theses sub layers are formed at different heights hence the maximum usable frequency, critical frequency and line of sight depends upon the height of a particular sub layer. Broadly it can be assumed that these layers are static over long periods but in reality it is not so, because the heights of these layers are not constant and the ion density is also variable due to dominant effects of sun. The electron density is totally depended on production and loss rate. The loss rate of any specific location can be dependent on various dynamic processes. With the development of science, it is possible for human being to communicate from one place to another place without using any physical connection. This is made possible in two ways, with and without help of satellite (with the help of sky wave propagation). The sky wave propagation is generally the preferred mode for HF communication links and is possible only with the help of ionospheric layer. The ionosphere provides long rang capabilities for commercial ship- to- shore communication and surveillance systems. The solar events widely called as space weather effects such as solar flares or coronal mass ejections can leads to worldwide communication blackouts on the HF range as well as on the trans-ionospheric communication system.

The behavior of ionosphere is also changes along with latitude therefore it is also divided into three regions i.e. high (polar), mid and low latitude. Low latitude region also consisting equatorial region, which has a very dynamic characteristic and also control the low latitudinal ionosphere. The equatorial ionosphere has some unique features such as equatorial ionization anomaly (EIA), the equatorial plasma fountain, equatorial electrojet etc. The magnetic field lines are approximately parallel over equatorial region and mutually perpendicular to the daytime eastward equatorial electric field, which gives rise to an upward drift of plasma at the equatorial region known as fountain effect. As the plasma is lifted to greater heights, it diffuses downwards along with the highly conducting geomagnetic field lines under the influence of gravity and pressure gradients. This gives rise to the formation of two crests on either side of the equator and the position of the crests is highly depended on the solar activity indices like F10.7 and Ap as well as on the other various parameters like neutral winds, meridonal wind etc.

The objective of this book is paying attention to study the short and long term variation over Antarctica. The quantitative understanding of high latitude ionosphere is still incomplete. It is needed to explore more therefore.

This book is describing about the space weather response of ionospheric layer and its characterization over the High latitude region, the F layer consists of one layer at night, but in the presence of sunlight, it divides into two layers, labeled F1 and F2. These F-layers are responsible for most sky wave propagation of radio waves, facilitating high frequency (HF, or shortwave) radio communications over long distances. They are thickest and most effective in refracting radio signals on the side of the earth facing the sun.

Space weather is influenced by the solar radiation and therefore the interplanetary magnetic flux (IMF) carried by the solar radiation plasma. a spread of physical phenomena are related to space weather, including geomagnetic storms and substorms, energization of the Van Allen radiation belts, ionospheric disturbances and scintillation of satellite-to-ground radio signals and long-range radar signals, aurora, and geomagnetically induced currents at surface. Coronal mass ejections (CMEs), their associated shock waves and coronal clouds also are important drivers of space weather as they will compress the magnetosphere and trigger geomagnetic storms. Solar energetic particles (SEP) accelerated by coronal mass ejections or solar flares can trigger solar particle events (SPEs), a critical driver of human impact space weather as they will damage electronics onboard spacecraft, and threaten the lives of astronauts also as increase radiation hazards to high-altitude, high-latitude aviation.

– Dr. Purushottam Bhawre

ACKNOWLEDGEMENTS

A formal statement of acknowledgment will hardly meet the end of justice while writing these words, I feel obliged to all of them who extended their inconceivable co-operation towards the achievements of whatever I have achieved.

I am extremely thankful to, Department of Physics, Barkatullah University, and Bhopal for providing all the necessary facilities and encouragement during the course of my work.

I wish to express my regards and deep sense of gratitude to **Director** and **Head of Radio and Atmospheric Science Division of National Physical Laboratory, (NPL) New Delhi,** provided the opportunity to participate in the XXVII Indian Scientific Expedition to Antarctica. I would like to express my sense of gratitude to **National Center for Antarctic and Ocean Research (NCPOR), Goa** for providing support for Antarctica expedition.

Special thanks to **Dr. Arun Chaturvedi** Leader 27th Indian Scintific Expedition to Antarctica for providing me timely motivation and encouragements for completing this work. I am tanksful to **John Turner,** British Antarctic Survey UK..

I express my special thanks to NOAA Space Environment Centre and NICT, Japan for providing the data used in this study. I am also thankful to OMNI Web Data Server for providing the data.

Finally, I would like to dedicate this Book to Indian base station "Maitri" Antarctica and National Physical Laboratory (NPL), New Delhi. In the end I can say only this much that all are not mentioned but none is forgotten.

– Purushottam Bhawre

CHAPTER I

ANTARCTICA INTRODUCTION

A NTARCTICA is that the highest, coldest and windiest continent. With very low amount of snowfall and practically no rainfall, most of the continent is technically a desert. The Antarctic icecap stores almost 70% of the world's freshwater and 90% of ice. The Southern Ocean surrounding this continent freezes up to a distance of 1500Km in winter, quite doubling the particular size (13.9 million sq km) of the continent. This is often also an area of atmospheric phenomenon and a continent where darkness prevails for months together during the Polar winters. The continent is as large as India and China put alongside no permanent habitation. There are not any forests, no perennial rivers and no industrial or military activity. There also are no market sor super bazaars. Explorers from different nations mingle with one-another and call themselves Polar men. Antarctica is that the ice-covered continent south of 60 degrees south latitude. Additionally to vast oceans, this region contains 14 million km2 of land, making it the fifth largest continent, nearly 1.5 times the dimensions of Canada. The bulk of the continent is roofed with ice, and glaciers line its coasts. The Antarctic is one among the harshest environments on the earth, a number of the strongest winds and lowest temperatures on earth are recorded within the Antarctic and snow and ice coverage are often up to four kilo meters thick.

The Antarctic is an environmentally significant region. It contains relatively untouched ecosystems that are scientifically valuable. The polar ice cap holds within it a record of past atmospheres that return tens or maybe many thousands of years. This enables an in depth study of the earth's natural climate cycles against which the importance of recent changes are often judged. The Antarctic is home to many unique and vulnerable wildlife species. Its marine environment sustains a good range of marine mammals, like seals and whales, at far greater levels than are found within the Arctic region. Short food chains make the Antarctic marine ecosystem very fragile and vulnerable to disruption.

While contributing to global biodiversity, the Antarctic also plays a central role within the world's ocean and climate systems. Approximately 80% of the worlds freshwater are frozen in Antarctic ice. The land mass and surrounding waters of the Antarctic provide essential nutrients to the remainder of the world's oceans and support life systems thousands of kilometers away. Fifth size larger among the world's continents. Its landmass is nearly wholly covered by a huge ice sheet.

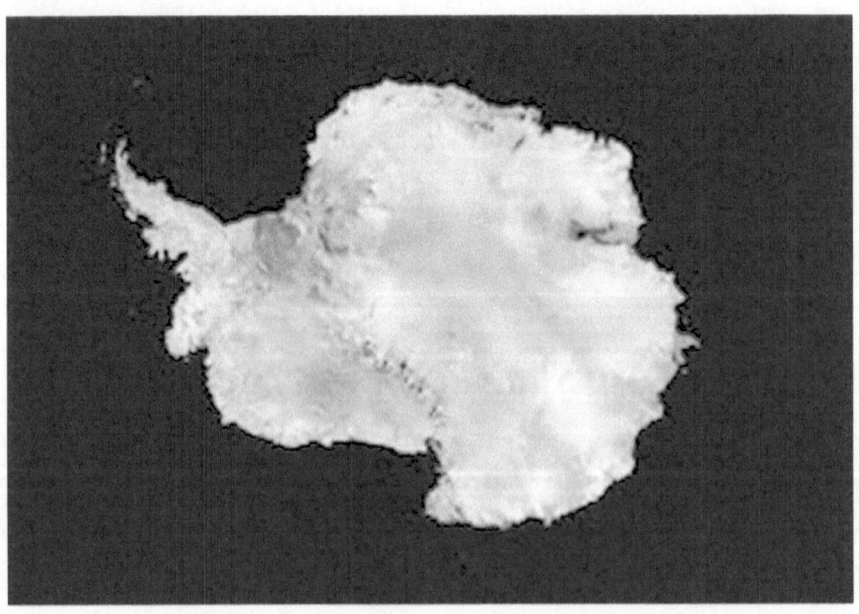

Antarctica continent

1.2 History of Antarctic

The history of Antarctica emerges from early Western theories of a huge continent, referred to as Terra Australis, believed to exist within the far south of the world. The term Antarctic, pertaining to the other of the Arctic Circle, was coined by Marinus of Tyre within the 2nd century AD. The rounding of the Cape of excellent Hope and Cape Horn within the 15th and 16th centuries proved thatTerra Australis Incognita ("Unknown Southern Land"), if it existed, was a continent in its title. In 1773 Cook and his crew crossed the Antarctic Circle for the primary time but although they found nearby islands, they didn't catch sight of Antarctica itself. It's believed he was as close as 150 mi (241.4 km) from the mainland. In 1820, several expeditions claimed to possess been the primary to possess sighted the shelf ice or the continent. A Russian expedition was led by Fabian Gottlieb von Bellingshausen and Mikhail Lazarev, a British expedition was captained by Edward Bransfield and an American sealer Nathaniel Palmer participated. The primary landing was probably just over a year later when American Captain Davys, a sealer, set foot on the ice. Several expeditions attempted to succeed in the South Pole within the early 20th century, during the 'Heroic Age of Antarctic Exploration'. Many resulted in injury and death. Norwegian Amundsen finally reached the Pole on December 14, 1911, following a dramatic race with the Englishman Robert Falcon Scott.

The continent may be a cold dry desert where access to water determines the abundance of life. While the terrestrial ecosystem contains quite thousand known species of organisms, most of those are microorganisms. Maritime Antarctica—the islands and coasts—supports more life than inland Antarctica, and therefore the surrounding ocean is as rich in life because the land is barren. From the late 18th to the mid-20th century, whalers and sealers plied the rich seas that surround the continent. Science then replaced whaling and sealing because the primary year-round act in ntarctica. Additionally, krill harvesting and other sorts of commercial fishing within the Southern Ocean expanded from the 1960s onwards. The new millennium saw tourism and (to a lesser extent) biological prospecting (the look for useful chemical compounds and genes in local species) become established sectors of the Antarctic economic landscape. Governments mandated many early expeditions—whether ostensibly economic, scientific, or exploratory in character to make territorial claims. With the International Geophysical Year (IGY) in 1957–58, this scale of scientific investigation of Antarctica began, and on December 1, 1959, the twelve countries that were active in Antarctica during the IGY signed the Antarctic Treaty. This treaty, which was an unprecedented landmark in diplomacy, preserves the continent for nonmilitary scientific pursuits and placed Antarctica under a world regime that, for the treaty's duration, holds all territorial claims in situ. The treaty bound its members indefinitely, with a review of its provisions possible after 30 years. A subsequent treaty, called the Madrid Protocol (adopted in 1991) prohibited mining, required environmental impact assessments for brand spanking new activities, and designated the continent as a natural reserve.

Knowledge about Antarctica has increased greatly since the IGY. Geologists, geophysicists, glaciologists, biologists, and other scientists have mapped and visited all of the continent's mountain regions. Until the 1970s, scientists relied on ground-based geophysical techniques like seismic surveys of the Antarctic ice sheets to reveal hidden mountain ranges and peaks. Advances in radar technology since then have resulted in airborne radio-echo sounding systems which will measure ice-thickness, which has enabled scientific teams to form systematic remote surveys of ice-buried terrains. Satellites and other remote-sensing technologies became key tools in providing mapping data.

The ice choked and stormy seas around Antarctica long hindered exploration by wooden-hulled ships. No lands break the relentless force of the prevailing west winds as they race clockwise round the continent, dragging westerly ocean currents along beneath. The southernmost parts of the Atlantic, Pacific, and Indian oceans converge into chilly oceanic water mass with unique biological and physical characteristics. Early penetration of this Southern (or Antarctic) Ocean within the look for fur seals led in 1820 to the invention of the continent. Icebreakers and aircraft now make access relatively easy, although still not without hazard in inclement conditions. Additionally, many tourists have visited Antarctica, which has underscored the worth of scenic resources within the continent's economic development.

The term Antarctic region refers to all or any area oceanic, island, and continental lying within the cold Antarctic zone south of the Antarctic Convergence, a crucial boundary around 55° S, with little seasonal variability, where warm subtropical waters meet and blend with cold polar waters (see also polar ecosystem). For legal purposes of the Antarctic Treaty, the arbitrary boundary of latitude 60° S is employed, south of which lies the Antarctic Treaty Area. The familiar map boundaries of the continent referred to as Antarctica, defined because the South Polar landmass and everyone its non-floating grounded ice, are subject to vary with current and future global climate change. The continent was ice-free during most of its lengthy geologic history, and there's no reason to believe it'll not become so again.

In the Western world, belief during a Cold Land a vast continent located within the far south of the world to "balance" out the northern lands of Europe, Asia and North Africa had existed for centuries. Aristotle had postulated symmetry of the world, which meant that there would be equally habitable lands south of the known world. The Greeks suggested that these two hemispheres, north and south, were divided by a "belt of fire", thanks to the overall observation that the climate got warmer and warmer the further south someone travelled, and no Europeans had gone past the equator to ascertain that this wasn't the case. It had been not until Prince Henry the avigator began in 1418 to encourage the penetration of the Torrid Zone within the effort to succeed in India by circumnavigating Africa that European exploration of the hemisphere began. In 1473 Portuguese navigator Lopes Gonçalves proved that the equator might be crossed, and cartographers and sailors began to assume the existence of another, temperate continent to the south of the known world. The doubling of the Cape of excellent Hope in 1487 by Bartolomeu Dias first brought explorers within touch of the Antarctic cold, and proved that there was an ocean separating Africa from any Antarctic land which may exist. Magellan, who skilled the Straits of Magellan in 1520, assumed that the islands of Tierra del Fuego to the south were an extension of this unknown southern land, and it appeared intrinsically on a map by Ortelius: Terra australis recenter inventa sed nondum plene cognita ("Southern land recently discovered but not yet fully known")

The obsession of the undiscovered continent culminated within the brain of Alexander Dalrymple, the brilliant and erratic hydrographer who was nominated by the Royal Society to command the Transit of Venus expedition to Tahiti in 1769. The command of the expedition was given by the admiralty to

Captain Cook. Sailing in 1772 with the Resolution, a vessel of 462 tons under his own command and therefore the Adventure of 336 tons under Captain Tobias Furneaux, Cook first searched vainly for Bouvet Island, then sailed for 20 degrees of longitude to the westward in latitude 58° S, then 30° eastward for the foremost part south of 60° S, a better southern latitude than had ever been voluntarily entered before by any vessel. On 17 January 1773 the Antarctic Circle was crossed for the primary time in history and therefore the two ships reached 67° 15' S by 39° 35' E, where their course was stopped by ice.

Cook then turned northward to seem for French Southern and Antarctic Lands, of the invention of which he had received news at Cape Town, but from the rough determination of his longitude by Kerguelen, Cook reached the assigned latitude 10° too Far East and didn't see it. He turned south again and was stopped by ice in 61° 52' S by 95° E and continued eastward nearly on the parallel of 60° S to 147° E. On 16 March, the approaching winter drove him northward for rest to New Zealand and therefore the tropical islands of the Pacific. In November 1773, Cook left New Zealand, having parted company with the journey, and reached 60° S by 177° W, whence he sailed eastward keeping as far south because the floating ice allowed. The Antarctic Circle was crossed on 20 December and Cook remained south of it for 3 days, being compelled after reaching 67° 31' S to face north again in 135° W.

A long detour to 47° 50' S served to point out that there was no land connection between New Zealand and Tierra del Fuego. Turning south again, Cook crossed the Antarctic Circle for the third time at 109° 30' W before his progress was once more blocked by ice four days later at 71° 10' S by 106° 54' W. now, reached on 30 January 1774, was the farthest south attained within the 18th century. With an excellent detour to the east, almost to the coast of South America, the expedition regained Tahiti for refreshment. In November 1774, Cook started from New Zealand and crossed the South Pacific without sighting land between 53° and 57° S to Tierra del Fuego; then, passing Cape Horn on 29 December, he rediscovered Roché Island renaming it Isle of Georgia, and discovered the South Hawaiian Islands (named Sandwich Land by him), the sole ice-clad land he had seen, before crossing the South Atlantic to the Cape of excellent Hope between 55° and 60°. He thereby laid open the way for future Antarctic exploration by exploding the parable of a habitable southern continent. Cook's most southerly discovery of land lay on the temperate side of the 60th parallel, and he convinced himself that if land lay farther south it had been practically inaccessible and of no value.

Admiral Fabian Gottlieb von Bellingshausen was one of the first to spot the continent of Antarctica.

It is certain that the expedition, led by von Bellingshausen and Lazarev on the ships Vostok and Mirny, reached on 28 January 1820 some extent within 32 km (20 mi) from Princess

Martha Coast and recorded the sight of an shelf ice at 69°21'28"S 2°14'50"W that became referred to as the Fimbul shelf ice. On 30 January 1820, Edward Brans field sighted Trinity Peninsula, the northernmost point of the Antarctic mainland. Von Bellingshausen's expedition also discovered Peter I Island and Alexander I Island, the primary islands to be discovered south of the circle.

The first landing on the Antarctic mainland is assumed to possess been made by the American Captain Davys, a sealer, who claimed to possess set foot there on 7 February 1821, though this is often not accepted by all historians.

In November 1820, Nathaniel Palmer, an American sealer trying to find seal breeding grounds, using maps made by the Loper whaling family, sighted what's now referred to as the Antarctic Peninsula, located between 55 and 80 degrees west. In 1823, James Weddell, a British sealer, sailed into what's now referred to as the Weddell Sea. Until the 20 th century, most expeditions were for commercial purpose, to seem for the prospects of seal and whale hunting. a bit of wood, from the South Shetland, was the primary fossil ever recorded from Antarctica, obtained during a personal us expedition during 1829-31, commanded by Captain Benjamin Pendleton.

Charles Wilkes, as commander of a us Navy expedition in 1840, discovered what's now referred to as Wilkes Land, a neighborhood of the continent around 120 degrees East.

After the North Magnetic Pole was located in 1831, explorers and scientists began trying to find the South Magnetic Pole. one among the explorers, James Clark Ross, a British military officer, identified its approximate location, but was unable to succeed in it on his 4 year-expedition from 1839 to 1843. Commanding British ships Erebus and Terror, he braved the Ice pack and approached what's now referred to as the Ross shelf ice, a huge floating shelf ice over 100 feet (30 m) high. His expedition sailed eastward along the southern Antarctic coast discovering mountains which were since named after his ships: Mount Erebus, the foremost active volcano on Antarctica, and Mount Terro.

The first documented landings on the mainland of East Antarctica was at Victoria Land by the American sealer Mercator Cooper on 26 January 1853.These explorers, despite their impressive contributions to South Polar exploration, were unable to penetrate the inside of the continent and, rather, formed a broken line of discovered lands along the coastline of Antarctica. Following the expedition South by the ships Erebus and Terror under James Clark Ross (January, 1841), he suggested that there have been no scientific discoveries, or 'problems', worth exploration within the far South. What followed is what historian H.R. Mill called 'the age of averted interest and within the following twenty years after Ross' return, there was a general lull internationally in Antarctic exploration.

The Southern Cross Expedition began in 1898 and lasted for 2 years. This was the primary expedition to overwinter on the Antarctic mainland (Cape Adare) and was the primary to form use of dogs and sledges. It made the primary ascent of the good Ice Barrier, (The Great Ice Barrier later became formally referred to as the Ross Ice Shelf). The expedition set a Farthest South record at 78°30'S. It also calculated the situation of the South Magnetic Pole.

The Discovery Expedition was then launched, from 1901–04 and was led by Robert Falcon Scott. It made the primary ascent of the Western Mountains in Victoria Land, and discovered the polar plateau. Its southern journey set a replacement Farthest South record, 82°17'S. Many other geographical features were discovered, mapped and named. This was the primary of several expeditions based in McMurdo Sound.

Expedition led by Robert Falcon Scott in 1901

A year later, the Scottish National Antarctic Expedition was launched, headed by William Speirs Bruce. 'Ormond House' was established as a meteorological observatory on Laurie Island within the South Orkneys and was the primary permanent base in Antarctica. The Weddell Sea was penetrated to 74°01'S, and therefore the coastline of Coats Land was discovered, defining the sea's eastern limits.

Ernest Shackleton, who had been a member of Scott's expedition, organized and led the Nimrod Expedition from 1907 to 1909. The expedition's primary objective was of reaching the South Pole. Based in McMurdo Sound, the expedition pioneered the Beardmore Glacier route to the South Pole, and therefore the (limited) use of motorised transport. Its southern march reached 88°23'S, a replacement Farthest South record 97 geographical miles from the Pole before having to show back. During the expedition, Shackleton was the primary to succeed in the polar plateau. Parties led by T. W. Edgeworth David also became the primary to climb Mount Erebus and to succeed in the South Magnetic Pole Expeditions from other countries. The First German Antarctic Expedition was sent to research eastern Antarctica in 1901. It discovered the coast of Wilhelm II II Land, and Mount Gauss. The expedition's ship became trapped in ice, however, which prevented more extensive exploration.

Erich von Drygalski led the First German Antarctic Expedition in 1901.

The Swedish Antarctic Expedition, operating at an equivalent time worked within the east coastal area of Graham Land, and was marooned on Snow Hill Island and Paulet Island within the Weddell Sea, after the sinking of its expedition ship. it had been rescued by the Argentinian naval vessel Uruguay. The French organized their first expedition in 1903 under the leadership of Jean-Baptiste Charcot. Originally intended as a relief expedition for the stranded Nordenskiöld party, the most work of this expedition was the mapping and charting of islands and therefore the western coasts of Graham Land, on the Antarctic Peninsula. a neighborhood of the coast was explored, and named Loubet Land after the President of France.A follow up trip was organized from 1908–1910 which continued the sooner work of the French expedition with a general exploration of the Bellingshausen Sea, and therefore the discovery of islands and other features, including Marguerite Bay, Charcot Island, Renaud Island, Mikkelsen Bay, Rothschild Island.

The Australasian Antarctic Expedition happened in 1911–1914 and was led by Sir Douglas Mawson. It is targeting the stretch of Antarctic coastline between Cape Adare and Mount Gauss, completing mapping and survey work on coastal and inland territories. Discoveries included Commonwealth Bay, Ninnis Glacier, Mertz Glacier, and Queen Mary Land. Major accomplishments were made in geology, glaciology and terrestrial biology.

The Imperial Trans-Antarctic Expedition of 1914–1917 was led by Ernest Shackleton and began to cross the continent via the South Pole. However, their ship, the Endurance, was trapped and crushed by Ice pack within the Weddell Sea before they were ready to land. The expedition members survived after a journey on sledges over Ice pack, a protracted drift on an ice-floe, and a voyage in three small boats to Elephant Island. Then Shackleton and five others crossed the Southern Ocean in an open boat called James Caird and made the primary crossing of South Georgia to boost the alarm at the whaling station Grytviken.

A related component of the Trans-Antarctic Expedition was the Ross Sea party, led by Aeneas Mackintosh. Its objective was to get depots across the good Ice Barrier, so as to provide Shackleton's party crossing from the Weddell Sea. All the specified depots were laid, but within the process three men, including the leader Mackintosh, lost their lives.

Shackleton's last expedition and therefore the one that brought the 'Heroic Age' to an in depth, was the Shackleton–Rowett Expedition from 1921–22 on board the ship Quest. Its vaguely defined objectives included coastal mapping, a possible continental circumnavigation, the investigation of sub-Antarctic islands, and oceanographic work. After Shackleton's death on 5 January 1922, Quest completed a shortened programme before returning home.

Women in Antarctica

Women were originally kept from exploring Antarctica until well into the 1950s. a couple of pioneering women visited the Antarctic land and waters before the 1950s and lots of women requested to travel on early expeditions, but were turned away. Early pioneers like Louise Séguin and Ingrid Christensen were a number of the primary women to ascertain Antarctic waters. Christensen was the primary woman to line foot on the mainland of Antarctica. the primary women to possess any fanfare about their Antarctic journeys were Caroline Mikkelsen who set foot on an island of Antarctica in 1935, and Jackie Ronne and Jennie Darlington who were the primary women to over-winter in Antarctica in 1947. the primary woman scientist to figure in Antarctica was Maria Klenova in 1956.[Silvia Morella de Palma was the primary woman to offer birth in Antarctica, delivering 3.4 kg (7 lb 8 oz) Emilio Palma at Argentina Esperanza base 7 January 1978. Women faced legal barriers and sexism that prevented most from visiting Antarctica and doing research until the late 1960s. The us Congress banned American women from traveling to Antarctica until 1969. Women were often excluded because it had been thought that they might not handle the acute temperatures or crisis situations. the primary woman from British Antarctic Survey to travel to Antarctica was Janet Thomson in 1983 who described the ban on women as a "rather improper segregation."

First women at the South Pole are Pam Young, Jean Pearson, Lois Jones, Eileen McSaveney, Kay Lindsay and Terry Tickhill.

Once women were allowed in Antarctica, they still had to fight against sexism and sexual harassment. However, a tipping point was reached in the mid-1990s when it became the new normal that women were part of Antarctic life. Women began to see a change as more and more women began working and researching in Antarctica

1.3 Life in a Cold Climate

I) Plants

Few land plants grow in Antarctica. this is often because Antarctica doesn't have much moisture (water), sunlight, good soil, or a heat. Plants usually only grow for a couple of weeks within the summer. However, moss, lichen and algae do grow. The foremost important organisms in Antarctica are the plankton which grow within the ocean.

II) Animals

One important source of food within the Antarctic is that the krill, which may be a general term for the tiny shrimp-like marine crustaceans. Krill are near rock bottom of the food chain: they prey on phytoplankton and to a lesser extent zooplankton. Krill are a food form suitable for the larger animals for whom krill makes up the most important a part of their diet. Whales, penguins, seals, and even a number of the birds that sleep in Antarctica, all depend upon krill. Whales are the most important animals within the ocean, and in Antarctica. They're mammals, not fish, meaning that they breathe air and don't lay eggs. Many various sorts of whales sleep in the oceans around Antarctica. Whalers have hunted whales for many years, for meat and blubber. Nowadays most whaling is completed within the Antarctic area.

Penguins only live south of the equator. Several different kinds sleep in and around Antarctica. the most important ones can stand nearly 4 feet (1.2m) tall and may weigh almost 100 pounds (40 kg). the littlest kinds are only about one foot (30 cm) tall. Penguins are large birds that swim alright but cannot fly. they need black backs and wings with white fronts. Their feathers are very tightly packed and make a thick cover. They even have a layer of woolly down under the feathers. The feathers themselves are coated with a kind of oil that creates them waterproof. A thick layer of blubber also keeps them warm. Penguins eat fish and are reception within the ocean. They are available abreast of the land or ice to get their eggs and lift the chicks. They nest together during a huge group.

III) Largest land animal

Since penguins are essentially marine animals, are there any entirely land animals on Antarctica? There are. the most important may be a wingless midge.

IV) People

People of the Antarctic live in there for a brief time to find out more about Antarctica, so most of the people that live there are scientists. Most are in national science stations on the coast. Some bases are far away from the ocean, for instance at the South Pole. They study the weather, animals, glaciers, and therefore the earth's atmosphere. Some scientists drill ice cores to seek out out about the weather way back. People that add the Antarctic must take care, because a blizzard can start anytime and anywhere, once they go distant from their shelter, they need to always take many food just just in case. Today people explore Antarctica using snowmobiles, which are faster than dogs and may pull heavier loads. Many come to Antarctica only for a brief visit, as a trip. There are companies in South America that have vacations to Antarctica, so people pay to travel there during a ship. Some people may take their own boats.

CHAPTER II

THE ROLE OF ANTARCTIC CLIMATE FOR THE GLOBAL CLIMATE SYSTEM

2.1 Introduction

The global climate system is driven by radiation, most of that, at any one time, arrives at low latitudes. Over the year as an entire the Equator receives concerning 5 times the maximum amount radiation because the poles, making an oversized Equator-to-pole temperature distinction. The region and oceanic circulations answer this huge horizontal gradient by transporting heat polewards (Trenberth and Caron, 2001). Really the climate system will be considered associate engine, with the low latitude areas being the warmth supply and also the Polar Regions the warmth sink. Though the dynamics of atmosphere ocean interaction area unit standard, the complexities of the warmth exchange engine and also the interactions of ice shelves and land ice with the ocean and atmosphere create prediction concerning global climate change a challenge.

The combination of tropical heating, poleward moving air and also the Coriolis force obligatory by the Earth's rotation ends up in the event of the Hadley Cell, associate region circulation during which air rises at the equator, making the tropical belt of air mass, and descends within the semitropics, forming the climatic zone high belt. At higher latitudes (60-65°S) the air ascends once more, making a unaggressive zone. The pressure gradient at the Earth's surface between the high within the semitropics and also the air mass at 60-65°S forces air to maneuver eastward underneath the influence of the Earth's rotation, making the mid-latitude westerlies that facilitate to drive surface waters east within the ACC. The air ascending at 60-65°S moves poleward at higher levels and sinks once more over the poles, forming a hard-hitting system over the Antarctica. The pressure gradient from the air mass at 60-65°S to the high over the continent offers rise to easterly winds on the Antarctic coast, driving the westward-directed Antarctic Coastal Current over the continental shelf; that shelf current tends to be targeted on the fronts of ice shelves (Deacon, 1937, Heywood, 2004). The most westward flow is that related to the Antarctic Slope Front, over the ocean floor. The north-to-south distribution of surface pressure around Antarctic continent is subject to exceptional variability within the intensity of the meridional pressure gradient and its zonal location. The circumpolar character of this variation it's known as the Southern annulated Mode (of variability) (SAM) (Trenberth and Jones 2007). The SAM may be seen as a live of the intensity of the westerly winds that propel the ACC. The SAM is that the hemisphere equivalent of the Arctic Oscillation (also referred to as the Northern annulated Mode) or the connected Atlantic Oscillation, that is measured from the pressure distinction between the island and Iceland. Variations within the SAM (which will be thought of as variations during a North - South pressure gradient) drive variability within the Southern Ocean's winds and currents: the vessel the gradient, the stronger the winds. Additionally to the SAM, there are a unit different vital modes of variability with meridional or zonal patterns. Each the atmosphere and ocean play major roles within the poleward transfer of warmth (Trenberth and Caron, 2001), with the atmosphere being chargeable for hr of the warmth transport, and also the ocean the remaining four-hundredth. Within the atmosphere, heat is transported by each unaggressive system and also the mean flow. Clockwise-circulating depressions carry heat air poleward on their Jap sides and cold air towards lower latitudes on their western flanks. The atmosphere is ready to reply comparatively quickly to changes within the high or low latitude heating rates, with storm tracks and also the mean flow dynamic on scales from days to years. The method by that air arrives at the poles conjointly imports pollutants from industrialized areas, the quantities area unit small compared with the Arctic, not least as a result of most industrialized area unit as are within the north, and also the mean flow is zonal around Antarctic continent instead of additional meridional as within the Arctic. The oceans carry heat and salt south towards the pole in higher ocean surface currents moving down the western sides of the Atlantic, Pacific and Indian Ocean basins (Schmitz, 1995, Lumpkin, 2007, Figure 2.1). additionally, heat and salt move south through the Atlantic within the sub-surface in Atlantic trouble, that rises to the surface close to the Antarctic coast (Figures 2.1 and 2.2), as a consequence of upwelling driven by the divergence of surface water forced north by the westerly winds of the Southern Ocean (Rintoul, 2001). Additionally to their transport by the large-scale currents, heat and salt ar transported by mesoscale eddies (with diameters of tens to many kilometers). within the Southern Ocean, wherever the meridional currents are commonly little, meridional transport by eddies is important (Diamond, 1981; Bryden, 1979; Hughes, 2001; Rintoul, 2001; Hogg, 2008). Eastward circulation is concentrated within the ACC, that lies broadly speaking between the Southern Boundary Front (SB) and also the Sub-Antarctic Front (SAF) (Rintoul, 2001). The core of the ACC lies roughly to a lower place the core of the predominant westerly winds. The ACC is an impressive feature within the international ocean's circulation. Stretching over a length of around 2000 km, it's the sole current to utterly encircle the world. It transports around one 140×106 m^3 of water per second (140 Sverdrups), creating it the world's largest stream. And it links the 3 main ocean basins (Atlantic, Pacific and Indian) into one international system by transporting heat and salt from one ocean to a different. The westerly winds of the Southern Ocean act on the surface waters, that ar forced north below the influence of the force of the Earth's rotation. Trouble from below wells up to switch these surface waters, as is clear from Figure 2.2, this ascent (Schmitz, 1995; Lumpkin, 2007) from the deep within the ACC forms the upward a part of the world over turning circulation. The northward moving surface water is cold, dense and sinks at the front to create Antarctic Intermediate Water (Wüst, 1935), that sinks northward at intermediate depths to permeate the word's oceans particularly within the hemisphere (McCartney, 1982). A number of the surface water reaches the SAF, wherever it sinks to create Sub-Antarctic Mode Water (Hanawa, 2001). Due to the excess of precipitation over evaporation at these latitudes, the surface water gets lighter as is moves north that explain however the Mode Water (lighter) involves overlie the Intermediate Water (heavier), despite their having a similar supply. South of the divergence zone, wherever upwelling takes place, the surface waters reaching coastal seas are cooled by contact with ice, and develop salt rejected once surface waters freeze to create ocean ice, thus losing buoyancy and changing into dense enough to sink. The sinking waters fall the sea bottom (Baines, 1998; Foldvik. 2004) and slope to create Antarctic Bottom Water, that spreads around Antarctic continent (Orsi, 1999) and aerates most of the world deep global ocean floor (Wüst, 1935; Hogg, 2001). These sinking cold waters give a reasonably uniform cold atmosphere for bottom abode (benthic) organisms.

Jets on the SAF and also the PF generally carry an oversized fraction of the transport of the ACC Figures 2.2 and 2.3; and choreographer et al., 2003), however different fronts will have comparable transports across individual oceanography sections. In fact, the ACC isn't one front however a posh system of fronts (e.g. Sokolov, 2002), many of that ar thought to be of circumpolar extent (Orsi, et al., 1995). This complicated system approach provides a brand new paradigm for considering the response of the ACC to global climate change

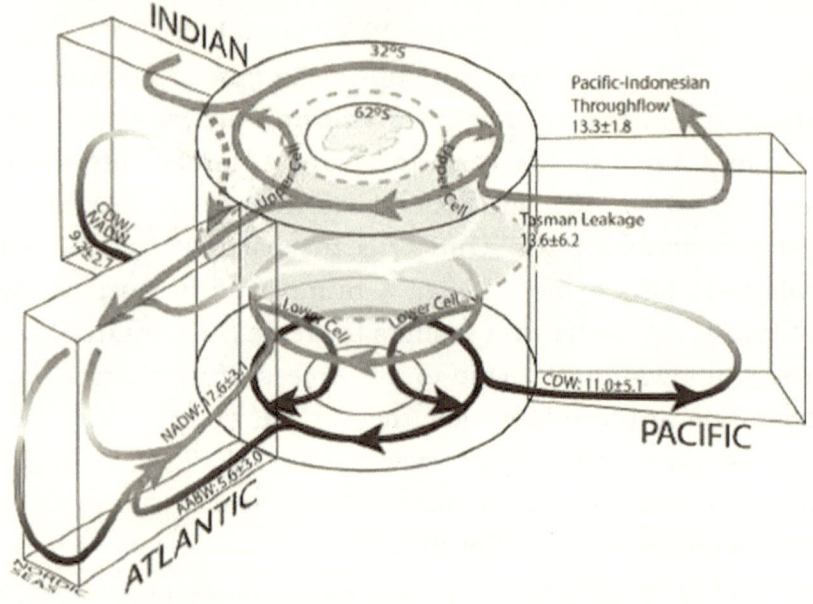

Figure 2.1 Model of the global ocean circulation, emphasising the central role played by the

Southern Ocean. From Lumpkin and Speer (2007). NADW = North Atlantic Deep Water; CDW = Circumpolar Deep Water; AABW = Antarctic Bottom Water. Units are in Sverdrups (1 Sv = 106 × m3 of water per second). The two primary overturning cells are the Upper Cell (red and yellow), and the Lower Cell (blue, green, yellow). The bottom water of the Lower Cell (blue) wells up and joins with the southward flowing deep water (green or yellow), which connects with the upper cell (yellow and red). This demonstrates the global link between Southern Ocean convection and bottom water formation and convective processes in the Northern Hemisphere.

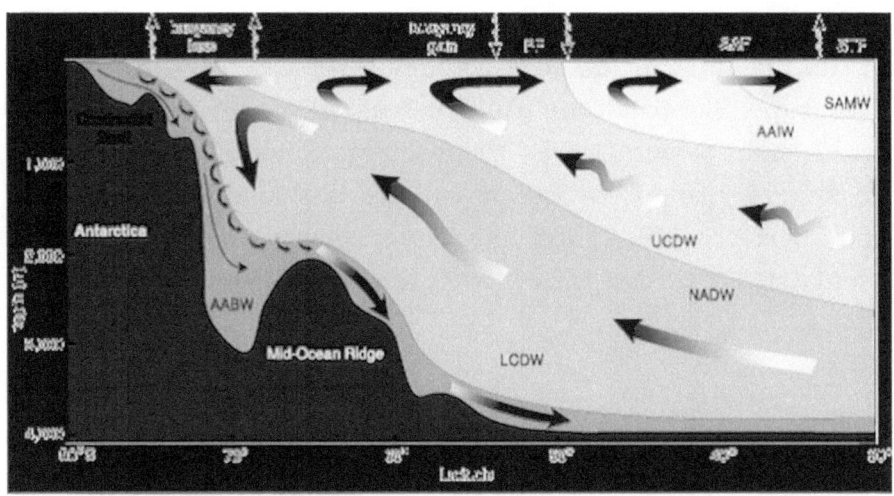

*Figure 2.2. South (left) to north (right) section through
the overturning circulation in the Southern Ocean. South-
flowing products of deep convection in the North Atlantic are
converted into upper-layer mode and intermediate waters
and deeper bottom waters and returned northward. Marked
are the positions of the main fronts (PF – Polar Front; SAF –
Sub-Antarctic Front; and STF – Subtropical Front), and water
masses (AABW – Antarctic Bottom Water; LCDW and UCDW,
Lower and Upper Circumpolar Deep Waters; NADW –North
Atlantic Deep Water; AAIW – Antarctic Intermediate Water and
SAMW – Sub-Antarctic Mode Water) (from Speer et al., 2000).
Note that as well as water moving north to south or vice versa,
it is also generally moving eastward (i.e. towards the observer
in the case of this cross section), except along the coast where
coastal currents move water westward (away from the observer).*

Superimposed on the circumpolar circulation system ar
regional dextrorotary gyres, principally the Weddell scroll and
Ross scroll, whose southern boundaries ar the west-moving Slope
Current, and whose outer limits reach to the east-moving ACC.
These gyres represent the elaborated pathway for transport to,
from, and on the continental margin. they're visible in ancient
oceanography surveys as dome-shaped structures encircled by
downward sloping plenty of equal density that reach towards
the coast within the south and towards the ACC within the
north. They are conjointly clearly discovered in Southern Ocean

numerical model outputs. Estimates of the extent of transport within the gyres ar few. The Weddell scroll carries thirty +/- ten Sverdrups (Sv) within the Weddell Sea (Fahrbach, 1994) and fifty six +/- eight Sv across the Greenwich Meridian (Klatt, 2005). The Ross scroll transports forty Sv across line of longitude 150°W, and therefore the Australian-Antarctic scroll seventy six +/- twenty six Sv across line of longitude 110°E (McCartney, 2007).

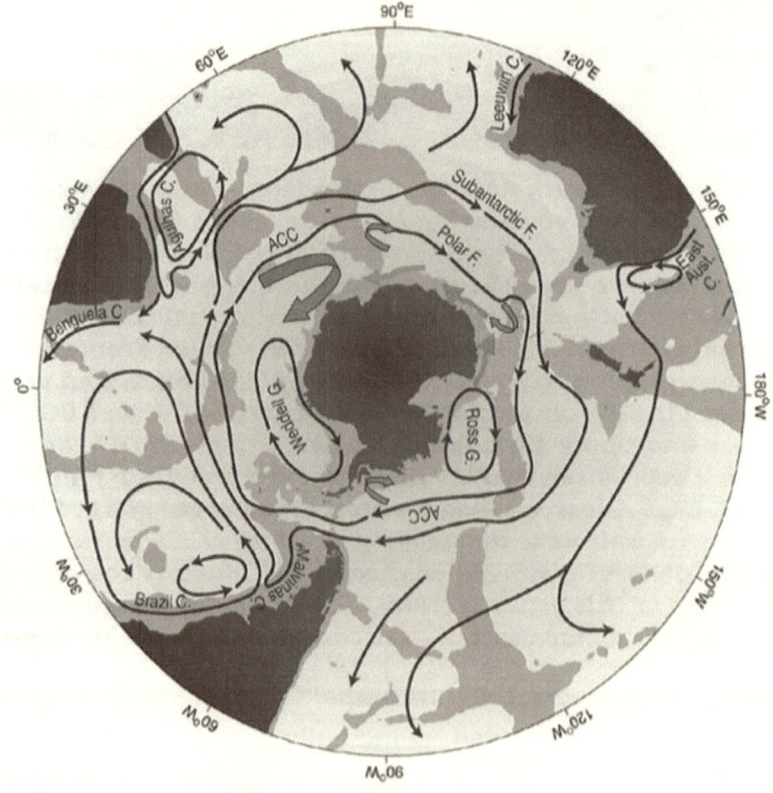

Figure 2.3. Schematic map of major currents south of 20°S (F = Front; C = Current; G =Gyre) (Rintoul et al, 2001); showing (i) the Polar Front and Sub-Antarctic Front, which are the major fronts of the Antarctic Circumpolar Current; (ii) Other regional currents; (iii) the Weddell and Ross Sea Gyres; and (iv) depths shallower than 3,500m shaded (all from Rintoul et al, 2001). In orange are shown (a) the cyclonic circulation west of the Kerguelen Plateau, (b) the Australian-Antarctic Gyre (south of Australia), (c) the slope current, and the (d) cyclonic circulation in the Bellingshausen Sea, as suggested

by recent modelling studies (Wang and Meredith, 2008), and observations – e.g. eastern Weddell Gyre - Prydz Bay Gyre (Smith et al, 1984), westward flow through Princess Elizabeth Trough (Heywood et al, 1999), and circulation east of Kerguelen Plateau (McCartney and Donohue, 2007).

The distribution of land and sea in the two polar regions is responsible for the very different atmospheric and oceanic circulations observed in each. Antarctica is a continent surrounded by ocean, while the Arctic is an ocean surrounded by land. The Antarctic continent is a sub-circular dome lying over the pole and surrounded by a broad swath of deep ocean, apart from the slight constriction of Drake Passage between the Antarctic Peninsula and South America (Fig. 2.3). The mean atmospheric flow and surface ocean currents are zonal in nature (parallel to latitude). The ACC, which is the major oceanographic feature of the Southern Ocean, flows unrestricted around the continent, isolating the cold high latitude land and sea areas from warm temperate mid and low latitude surface waters. This has only been the case since about 30 million years ago when the Drake Passage opened up, the ocean currents had more of a meridional component (parallel to longitude), which allowed greater poleward penetration of warm waters from temperate latitudes. In contrast, as mentioned earlier, surface ocean circulation in the Arctic tends to be meridional, especially the northwards flow of warm water at the surface in the North Atlantic Current. The Southern Ocean plays a key role in the global carbon cycle. The upwelling deep water south of the Polar Front brings to the surface dissolved nutrients and carbon dioxide (CO_2), and releases this gas to the atmosphere. In contrast, the Intermediate Water and Mode Water masses sinking north of the Polar Front (Figure 2.2) take up CO_2 from the atmosphere. These complementary processes make the Southern Ocean both a source and a sink for atmospheric CO_2 (see Figure 2.4; from Sabine, 2004).

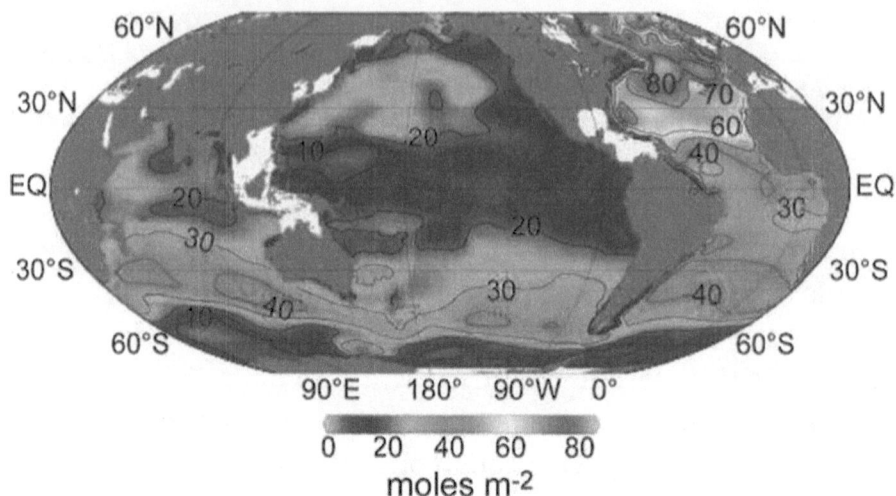

Figure 2.4. Column inventories of anthropogenic CO2 in the ocean. Dissolved old CO2 is lost to the atmosphere south of the Polar Front, where NADW wells up to the surface close to the coast (purplish colours); most of this old CO2 is not anthropogenic. The cold sinking Mode and Intermediate Waters north of the polar front (green colours) have extracted considerable anthropogenic CO2 from the atmosphere (from Sabine et al., 2004).

Because of its upwelling nutrients, the Southern Ocean is that the world's most biologically productive ocean. Nonetheless, its productivity is restricted by the low availability of micronutrients like iron, except round the islands that are scattered through the ACC. As a result the Southern Ocean is assessed as High Nutrient Low Chlorophyll (HNLC). Through photosynthesis, the expansion of phytoplankton extracts CO2 from the atmosphere and pumps it to the seabed or into subsurface waters through the sinking of decaying organic matter. Without this process, and without the answer of CO2 in cold dense sinking water near the coast, the buildup of CO2 within the atmosphere would be much faster.

In the summer period, the sun is above the horizon for long durstion, the Antarctic receives more radiation than the tropics, but the highly reflective ice and snow covered surfaces reflect much of this radiation back to space. This reflection is one among the important feedback mechanisms found within the ice covered polar regions, because it enhances cooling. Where snow melts, exposing large patches of bare (dark) ground, or where sea ice melts exposing ice free (dark) ocean, radiation are going to be absorbed instead of reflected, and therefore the environment will warm. The Polar Plateau of East Antarctica experiences very low temperatures due to its high elevation, the shortage of cloud and water vapor within the atmosphere, and therefore the isolation of the region from the relatively warm maritime air masses found over the Southern Ocean. The very cold temperatures within the interior of Antarctica, including its isolation from warm, moist air masses, mean that precipitation there's very low, with only about 5 cm water equivalent falling per annum (King and Turner, 1997). That creates much of Antarctica a desert and therefore the driest continent on Earth. The low temperatures mean that there's little or no evaporation and sublimation, in order that although the quantity of precipitation is little, it builds up year by year to make the ice sheet. Many blizzards tend to be fallen snow resuspended, instead of new snow.

Temperatures are much less extreme within the Antarctic coastal region than on the plateau generally, at the most of the coastal stations the monthly mean summer temperatures never rise above melting point, although daily temperatures may show excursions above freezing in places during summer. There are exceptions. the very best temperatures on the continent are found on the western side of the Antarctic Peninsula where there's a prevailing northwesterly wind; there temperatures can rise to many degrees above freezing during the summer, and monthly means are positive for 2-4 months of the year. Temperatures also tend to be above freezing sometimes in places just like the Schirmacher Oasis near the coast in Dronning Maud Land, where there are ice-free lakes within the austral summer. due to the shortage of incoming radiation, the Antarctic stratosphere in winter is extremely cold. a robust gradient develops between the continent and midlatitudes (Figure2.5), isolating a pool of very cold air above Antarctica. Very strong winds develop along this thermal gradient. They are stronger than the equivalent winds found within the Arctic, because the Equator-to-pole temperature difference is larger within the south. The pool of cold air and its strong surrounding winds together form the polar vortex. It plays a crucial part in determining the atmospheric circulation of the high southern latitudes, also as within the formation of the hole, where the polar vortex acts as a 'containment vessel' allowing chlorofluorocarbon compounds (CFCs) to create up during the winter.

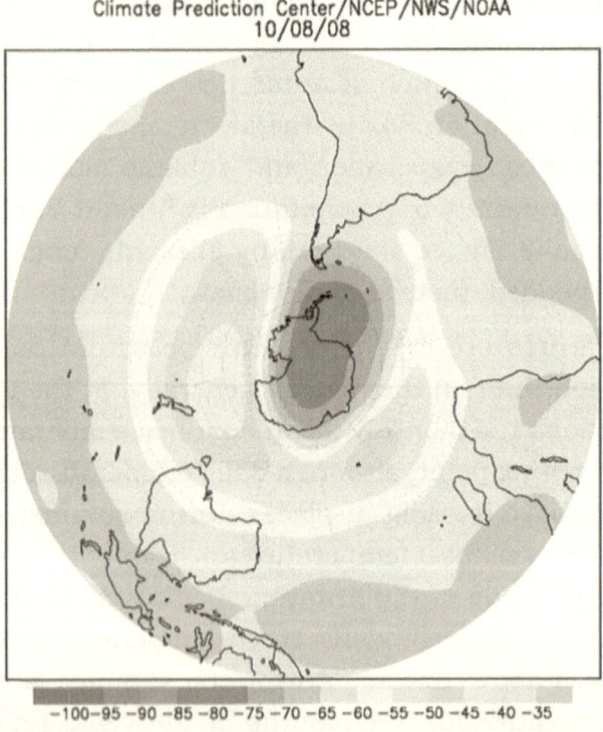

50MB SH Temperature Analysis
Climate Prediction Center/NCEP/NWS/NOAA
10/08/08

−100−95 −90 −85 −80 −75 −70 −65 −60 −55 −50 −45 −40 −35

Figure 2.5 The polar vortex above Antarctica (indicated by the red colours) in mid-winter (August) as seen via the temperatures in degrees C on the 50 hPa pressure surface (roughly 20 km elevation above mean sea level). Generated from the NCEP reanalysis using their online, interactive chart drawing system at: www.cdc.noaa.gov/data/reanalysis/reanalysis.shtml.

The Antarctic climate system varies from time to and is closely coupled to other parts of the worldwide climate system. On the longest time scales it fluctuates on Milankovitch frequencies (20 ka, 41 ka, 100 ka) in response to variations within the Earth's orbit round the sun that cause regular variations within the Earth's climate on these recurrent time scales. Proxy data from ice cores show that since the Last Glacial Maximum (LGM) at about 21 ka before present (BP) there are variety of climatic fluctuations across the continent. One among the foremost marked was the Mid Holocene warm period, which is present in various records from Antarctica. Ice cores also reveal shifts within the intensity of the atmospheric circulation, with the westerlies weakening at 5,200-5,400 years ago and strengthening around 1,200 years ago. Because long time-series of observations didn't begin until the IGY (1957-58), the instrumental period in Antarctic is merely about 50 years long (measurements started before then but weren't continuous). Since proxy data show oscillations on longer time scales than 50 years, it's accepted that the instrumental period only provides a snapshot of change within the Antarctic.

Reliable weather charts for top southern latitudes have only been available since the late 1970s. They reveal an excellent deal about the patterns within the atmospheric circulation of high southern latitudes. Over the previous couple of decades there has been a marked drop by pressure round the Antarctic coast and a rise in mid-latitudes. The rise within the pressure gradient across the Southern Ocean has strengthened the surface winds, which successively has affected ocean currents and therefore the distribution of sea ice. Studies of coupled ocean-atmosphere general circulation models show that the strength of the SAM should increase because the Earth warms, confirming what's known from observations. But the strength of the SAM is additionally modulated by the hole over Antarctica.

Stratospheric ozone is a crucial constituent of the upper atmosphere above the Antarctic, where ozone levels began to say no within the late 1970s following widespread releases of CFCs and halons into the atmosphere. We now know that the presence of CFCs within the Antarctic stratosphere leads to a posh reaction during the spring that destroys virtually all ozone at altitudes between 14 and 22 km, especially within the polar vortex where temperatures are coldest. The depletion in ozone maintains the cooling within the polar vortex, which accentuates the winds round the vortex. This strengthening of the winds at high altitude in spring then propagates downwards within the atmosphere through time, strengthening surface winds during the summer and autumn. Thus one effect of the hole has been to intensify the SAM signal, further strengthening the westerly winds round the Antarctic during the summer and autumn. within the strengthened SAM, with its stronger winds, we see two samples of changes from outside the region having a profound impact on the Antarctic environment. One impact comes from the CFCs liable for the 'ozone hole', and therefore the other comes from greenhouse emission emissions liable for heating. Both CFCs and greenhouse gases have mainly been released within the hemisphere during the economic era. Both have had, and still have, a profound effect on the radiation balance of the Antarctic atmosphere. Although the Antarctic is way far away from where most of the solar power enters the system at tropical latitudes, it's still influenced by variability in tropical conditions, and signals of low latitude climate variability are often identified in Antarctica and therefore the Southern Ocean (Turner, 2004). Additionally, there's increasing evidence that signals also can be transmitted within the other way from high to low latitudes.

Statistically significant long-range linkages between high and low latitudes pass the name of teleconnections. They will happen via the atmosphere and/or the ocean, although the timescales are usually rather different. The foremost rapid teleconnections generally occur via the atmosphere, with storm track changes occurring on the size of days or weeks. The El Niño-Southern Oscillation (ENSO) is one among the most important climatic cycles on Earth, working on the size of years to decades. It's its origins within the tropical Pacific, but its effects are often felt across the planet. As discussed during a number of places during this volume, ENSO signals are often identified within the physical and biological environment of the Antarctic, although a number of the links aren't robust and there are often large differences within the extra-tropical response to near-identical events within the tropics. The circulation of the upper layers of the ocean can change over months to years, but the deep ocean and therefore the global thermohaline circulation (THC) require decades to centuries to reply. At the opposite extreme, fast wave propagation within the ocean takes place on timescales of just a couple of days.

About 30 years of reliable atmospheric reanalysis outputs are available for the high southern latitudes, so it's been possible to determine the character of the broad-scale teleconnections typical of the hemisphere. There's evidence of decadal timescale variability in a number of these linkages, but with such a brief data set it's impossible at the present to realize insight into how the teleconnections may vary on longer timescales. High resolution ice core records collected from areas where the speed of precipitation of snow is high shed some light on teleconnections over the century timescale, and where the speed of accumulation is high enough they will even give seasonal data, which is vital since some teleconnections are only present in individual seasons. Through most of the Holocene (roughly the past 12,000 years) hemisphere events have lagged behind hemisphere ones by several hundred years. That has changed in recent decades, and therefore the hemisphere signal of rising temperature since about 1850 AD has paralleled the hemisphere one. natural process within the two hemispheres now appears to be synchronous a radical departure from former times, which suggests a replacement and different forcing, presumably regarding anthropogenic activity within the sort of enhanced greenhouse gases. Comparing data on winds and temperatures from northern and hemisphere ice cores confirms that the wind/temperature fields of today differ from those of the recent past indicating that the fashionable day atmosphere isn't an analogue for that of the so called Medieval Warm Period. they vary from the glaringly obvious, just like the collapse of the Larsen B shelf ice in February-March 2002, and therefore the warming of varied parts of the Antarctic Peninsula over the past 50 years (King, 1994), to the rather more subtle, just like the 0.2°C warming of the Southern Ocean, which - though alittle amount represents a serious transfer of warmth when summed over a huge area.

2.2. Antarctica Metrology

Weather preparedness is usually a key consideration in Antarctica, whether you're flying a plane, fixing a roof, drilling ice cores or counting penguins.

The Australian Bureau of Meteorology features a long history of supporting the safe and efficient undertaking of the Australian Antarctic program. We've been present at each of Australia's Antarctic and sub-Antarctic stations since they were established. Bureau meteorologists were also a part of a number of the sooner heroic age expeditions, including Griffith Taylor, who represented the 'Weather Service' on Scott's Terra Nova Expedition (1910–1913), and Bureau officer George Ainsworth, who led Douglas Mawson's team on Macquarie Island in 1911.

Safer and more efficient operations aren't the sole reason the Bureau partners with the Australian Antarctic Division. The Bureau also has the important Government mandate to watch the Australian climate. The Bureau's continuous, quality managed and long-standing record of weather observations from the four stations, provides Australians and therefore the global community with information to research and monitor Antarctica's weather and climate, and help understand Antarctica's role within the larger Earth system.

Because weather and climate don't recognise geopolitical boundaries, most nations have agreed to collaborate within the taking and sharing of weather observations for the advantage of the worldwide community. These efforts are coordinated at the worldwide level by the planet Meteorological Organisation (WMO), which today comprises 185 member nations, and during which the Australian Bureau of Meteorology plays a proactive role. The Bureau's weather observation and climate monitor program follows strict WMO standards, to manually and automatically record and distribute information on wind, air pressure, humidity, temperature, radiation and space weather.

As with most things Antarctic, we partner with other organisations. We enjoy the Australian Antarctic Division's remote Automatic meteorological observation post network and VHF wind-profiler for the supply of our aviation forecast services and for the initialisation of our global weather model, the Australian Community Climate and Earth System Simulator (ACCESS). We also partner with the CSIRO and therefore the Australian Nuclear Science and Technology Organisation.

We support various national and international research campaigns. These include the present Macquarie Island Cloud and Radiation Experiment (Australian Antarctic Magazine 30: 13, 2016), the upcoming Antarctic Cloud and Radiation Experiment, and two wider-ranging Southern Ocean cloud and radiation experiments which will use instruments ashore, ships, aircraft and satellites. Other important research is being undertaken in Hobart to guage the skill of our seasonal sea-ice forecasts from ACCESS-S ('S' for seasonal) also as evaluate the skill and optimise the polar physics of our shorter term ACCESS-G (global) forecasts in Antarctica.

In terms of direct service to the Antarctic Division, time has proven that meteorologists perform best after they need experienced the local environment first-hand, which they convey most effectively when embedded within the operation. This is often particularly salient within the Australian context, where many forecasters have focused their training and knowledge on phenomena like tropical cyclones, fire weather and flood forecasting, instead of the polar regions. Until they undertake their one month of pre departure polar meteorological training in Hobart, most of our recruited Antarctic forecasters would haven't considered the nuances of katabatic interactions with transient low systems which will whomp up 160 km/h winds in minutes, or ocean-air-ice interactions which will see a ship disappear in its own localised fog patch.

Our forecasters are typically deployed to where aviation operations are focussed, like at Casey and Davis over summer and at Macquarie Island during the station resupply. we've an annually recruited workforce of 11 over-wintering observing and technical staff, five summer weather forecasters (often including a Royal Australian Navy forecaster), and another officer or two on project work on any given time. Over the quieter winter season, forecast support is provided on a by-request basis from Hobart. additionally to marine and atmospheric weather, we also warn for tsunami, abnormally high tides, and space weather (which can have detrimental effects on satellite systems and communications).

Antarctic climate system isn't yet fully understood. Timo Vihma is involved in a world scientific research called Antarctic Meteorology and its Interaction with the Cryosphere and Ocean, which is aimed toward increasing our understanding of the interactions happening between the atmosphere, ice and oceans. The research provides the inspiration for the modelling of interaction phenomena, and it'll also contribute to the event of weather outlook and climate models. Timo Vihma may be a meteorologist, but he features a strong background in marine research. He has been studying the dynamics of sea ice and therefore the interactions between the ocean and therefore the atmosphere since the 1990

2.3 Physical Climate

The unique weather and climate of Antarctica provide the idea for its familiar appellations—Home of the Blizzard and White Desert. far and away the coldest continent, Antarctica has winter temperatures that range from −128.6 °F (−89.2 °C), the world's lowest recorded temperature, measured at Vostok Station (Russia) on July 21, 1983, on the high inland ice sheet to −76 °F (−60 °C) near water level. Temperatures vary greatly from place to put, but direct measurements in most places are generally available just for summertime. Only at fixed stations operated since the IGY have year-round measurements been made. Winter temperatures rarely reach as high as 52 °F (11 °C) on the northern Antarctic Peninsula, which, due to its maritime influences, is that the warmest a part of the continent. Mean temperatures of the coldest months are −4 to −22 °F (−20 to −30 °C) on the coast and −40 to −94 °F (−40 to −70 °C) within the interior, the coldest period on the polar plateau being usually in late August just before the return of the sun. Whereas midsummer temperatures may reach as high as 59 °F (15 °C) on the Antarctic Peninsula, those elsewhere are usually much lower, starting from a mean of about 32 °F (0 °C) on the coast to between −4 and −31 °F (−20 and −35 °C) within the interior. These temperatures are far less than those of the Arctic, where monthly means range only from about 32 °F in summer to −31 °F in winter.

2.3.1 Radiations

Antarctica's ice sheets are still releasing radioactive chlorine from marine nuclear weapons tests within the 1950s, a replacement study finds. this means regions in Antarctica store and vent the radioactive element differently than previously thought. The results also improve scientists' ability to use chlorine to find out more about Earth's atmosphere.

Scientists commonly use the radioactive isotopes chlorine-36 and beryllium-10 to work out the ages of ice in ice cores, which are barrels of ice obtained by drilling into ice sheets. Chlorine-36 may be a present radioactive isotope, meaning it's a special mass than regular chlorine. Some chlorine-36 forms naturally when argon gas reacts with cosmic rays in Earth's atmosphere, but it also can be produced during nuclear explosions when neutrons react with chlorine in seawater.

Nuclear weapons tests within the us administered within the Pacific during the 1950s and therefore the 1960s caused reactions that generated high concentrations of isotopes like chlorine-36. The radioactive isotope reached the stratosphere, where it traveled round the globe. a number of the gas made it to Antarctica, where it had been deposited on Antarctica's ice and has remained ever since.

Other isotopes produced by marine nuclear bomb testing have mostly returned to pre-bomb levels in recent years. Scientists expected chlorine-36 from the nuclear bomb tests to possess also rebounded. But new research in AGU's Journal of Geophysical Research: Atmospheres finds the Vostok region of Antarctica is constant to release radioactive chlorine into the atmosphere. Since naturally produced chlorine-36 is stored permanently in layers of Antarctica's snow, the results indicate the location surprisingly still has manmade chlorine produced by bomb tests within the 1950s and within the 1960s.

"There is not any more nuclear chlorine-36 within the global atmosphere. That is... why we should always observe natural chlorine-36 levels everywhere," said Mélanie Baroni, a geoscientist at the Centre for Research and Teaching in Geosciences and therefore the Environment in Aix-en-Provence, France, and co-author of the new study.

Studying the chlorine's behavior in Antarctica can improve ice dating technology, helping scientists better understand how Earth's climate evolved over time, consistent with the study's authors.

2.3.2 Temperature

The highest temperature ever recorded on Antarctica was 20.75 °C (69.3 °F) at Comandante Ferraz Antarctic Station on 9 February 2020, beating the previous record of 18.3 °C (64.9 °F) at Esperanza Base, on the northern tip of the Antarctic Peninsula, on 6 February 2020, a better temperature of 19.8 °C (67.6 °F) recorded at Signy Research Station on 30 January 1982 was the record for the Antarctic region encompassing all land and ice south of 60° S. The Antarctic temperature changes during the last several glacial and interglacial cycles of this glacial period rock bottom air temperature record, rock bottom reliably measured temperature on Antarctica was assail 21 July 1983, when −89.2 °C (−128.6 °F) was observed at Vostok Station. For comparison, this is often 10.7 °C (19.3 °F) colder than subliming solid (at water level pressure). The altitude of the situation is 3,488 meters (11,444 feet). Satellite measurements have identified even lower ground temperatures, with −93.2 °C (−135.8 °F) having been observed at the cloud-free East Antarctic Plateau on 10 August 2010.[6]The lowest recorded temperature of any location on surface at 81.8°S 63.5°E was revised with new data in 2018 in nearly 100 locations, starting from −93.2 °C (−135.8 °F)[7] to −98 °C (−144.4 °F). This unnamed a part of the Antarctic plateau, between Dome A and Dome F, was measured on 10 August 2010, and therefore the temperature was deduced from radiance measured by the Landsat 8 and other satellites, and discovered during a National Snow and Ice Data Center review of stored data in December 2013 but revise by researcher on 25 June 2018. This temperature isn't directly like the –89.2 °C reading quoted above, since it's a skin temperature deduced from satellite-measured upwelling radiance, instead of a thermometer-measured temperature of the air 1.5 m (4.9 ft) above the bottom surface.

The mean annual temperature of the inside is –57 °C (–70.6 °F). The coast is warmer; on the coast Antarctic average temperatures are around –10 °C (14.0 °F) (in the warmest parts of Antarctica) and within the elevated inland they average about –55 °C (–67.0 °F) in Vostok. Monthly means at McMurdo Station range from –26 °C (–14.8 °F) in August to –3 °C (26.6 °F) in January. At the South Pole, the very best temperature ever recorded was –12.3 °C (9.9 °F) on 25 December 201. Along the Antarctic Peninsula, temperatures as high as 15 °C (59 °F) are recorded, clarification needed] though the summer temperature is below 0 °C (32 °F) most of the time. Severe low temperatures vary with latitude, elevation, and distance from the ocean. East Antarctica is colder than West Antarctica due to its higher elevation.[citation needed] The Antarctic Peninsula has the foremost moderate climate. Higher temperatures occur in January along the coast and average slightly below freezing.

I) Surface temperature

Surface temperature trends across the Antarctic are often determined employing a number of various sorts of data, including the in-situ observations, satellite infra-red imagery and ice core isotope measurements. so as to urge an inexpensive estimate of trends it's necessary to use of these data. The in-situ observational record of Antarctic surface temperatures is quite sparse and sporadic before the IGY, although the Orcadas series from Laurie Island, South Orkney Islands began in 1903 and therefore the Faraday Station/Argentine Islands record began in 1947. However, we are fortunate in having around 16 stations on the Antarctica or islands that have reported on a near-continuous basis since the IGY. additionally, an extra six stations started reporting during the 1960s, in order that we've around twenty-four statistic that allow the investigation of temperature trends. Unfortunately, the overwhelming majority of the stations are within the Antarctic coastal region or on the islands of the Southern Ocean, with only Vostok and Amundsen-Scott Station being within the interior of the continent. The in-situ record has been employed by several workers to research temperature changes across the continent and Southern Ocean (Jacka, 1991; Jack, 1998; Jones, 1995; Raper, 1984). Many of the records were scattered across variety of knowledge centres and it had been unclear on the quantity of internal control that had been administered on the observations. SCAR therefore initiated the READER (Reference Antarctic Data for Environmental Research) project to bring as many of the observations together as possible, internal control the info and produce a replacement data base of monthly mean temperatures (Turner, 2004). The READER data base is out there online at http://www.antarctica. ac.uk/met/READER/.The READER data base has now been utilized in variety of studies concerned with the climate of the Antarctic, including that of Turner. (2005a), which considered changes since the beginning of the routine instrumental record. Here we'll use the READER data base and therefore the online meteorological data maintained by Dr. Gareth Marshall (http:// www.antarctica.ac.uk/met/gjma/) to look at how Antarctic temperatures have changed over the amount of the instrumental record.

Surface temperature trends from the station data since the first 1950s illustrate a robust dipole of change, with significant warming across the Antarctic Peninsula, but with little change across the remainder of the continent (Figure 4.8a). the most important warming trends within the annual mean data are found on the western and northern parts of the Antarctic Peninsula. Here Faraday/Vernadsky Station has experienced the most important statistically significant (<5% level) trend of +0.53 o C/dec for the amount 1951-2006. Rothera station, some 300 km to the south of Faraday, has experienced a bigger annual warming trend, but the shortness of the record and therefore the large inter-annual variability of the temperatures means the trend isn't statistically significant. Although the region of marked warming extends from the southern a part of the western Antarctic Peninsula north to the South Shetland, the speed of warming decreases faraway from Faraday/Vernadsky, with the long record from Orcadas on Laurie Island, South Orkney Islands only having experienced a warming of +0.20oC/decade. This record covers a 100-year period instead of the 50 years for Faraday. For the amount 1951-2000 the temperature trend was +0.13oC/decade. Determining temperature trends across the inside of the Antarctic is difficult as there are only two stations with long records. However, attempts are made to extrapolate the station trends across the remainder of the continent. Chapman (2007) produced estimates of annual tends and located the best warming over the Antarctic Peninsula, but with alittle warming (~0.1° C/dec) across West Antarctica and far of East Antarctica. However, they also found cooling during a swath from the South Pole to Halley Station. Steig. (2009) use statistical climate-field-reconstruction techniques to supply similar fields of trends for the seasons and therefore the year as an entire. The annual trends significant warming over most of West Antarctica with trends greater than 0.1° C/dec over the last

50 years. The trends are greatest during the winter and spring. There has been an excellent deal of debate about the causes of the recent temperature changes across the continent. The summer warming on the eastern side of the Antarctic Peninsula has been shown to be a result of anthropogenic activity, and particularly the spring time loss of stratospheric ozone (Marshall, 2006). For the continent as an entire Gillett (2009) administered a proper attribution study to work out whether the observed changes were within the range of natural climate variability or whether or not they were a results of anthropogenic forcing. They found that recent changes weren't according to internal climate variability or natural climate drivers alone, and were directly due to human influence.

Prior to the establishment of the research stations within the middle of the 20 th Century we are reliant on ice core data to research surface temperature changes. Many studies have used single and multiple cores to research changes at selected sites or to research regional change. However, 'stacking' multiple cores can provide insight into Antarctic-wide change. Schneider (2006) stacked several isotope records from ice cores to get a continental pattern of temperature over the past 200 years.

II) Upper air temperature changes

Analysis of Antarctic radiosonde temperature profiles indicates that there has been a warming of the troposphere and cooling of the stratosphere over the last 30 years. this is often the pattern of change that might be expected from increasing greenhouse gases, however, the midtropospheric warming in winter is that the largest on Earth at this level. the info show that regional mid-tropospheric temperatures have increased most round the 500 hPa level with statistically significant changes of 0.5 – 0.7°C/decade (Figure 2.6) (Turner et al., 2006). Figure 2.6 indicates warming at many of the radiosonde stations round the continent, but a transparent pattern of winter warming is clear round the coast of East Antarctica and at the pole.

The warming is represented within the ECMWF 40 year reanalysis, which isn't surprising since the radiosonde ascents were assimilated into the system. after all the warming trends are slightly larger than when computed from the radiosonde data, since there's a known slight cold bias within the early a part of the reanalysis data set.

The exact reason for such an outsized mid-tropospheric warming isn't known at the present. However, it's recently been suggested that it's going to, a minimum of partially, be a results of greater amounts of polar stratospheric cloud (PSC) during the winter. PSCs are a feature of the cold Antarctic winter, forming at temperatures below about -78° C. However, the Antarctic stratosphere has cooled in recent decades because the greenhouse emission ozone is now missing from the lower stratosphere in spring, and therefore the greenhouse emission CO_2 is concentrated within the troposphere and results in further cooling of the stratosphere. Analysis of stratospheric temperatures within the reanalysis data sets suggest that over the last 30 years the world where PSCs might form in winter has increased in size, so promoting the formation of more PSCs. Once present, PSCs act like all other cloud, giving a warming below their level and cooling above. we've little data on the optical properties of PSCs, but modelling suggests that if the optical depth within the infrared is around 0.5 then a greater amount of PSCs could provides a mid-tropospheric warming. PSCs aren't currently represented explicitly in climate models, but if further research shows that they're liable for the massive winter season mid-tropospheric warming they have to be represented more realistically within the models.

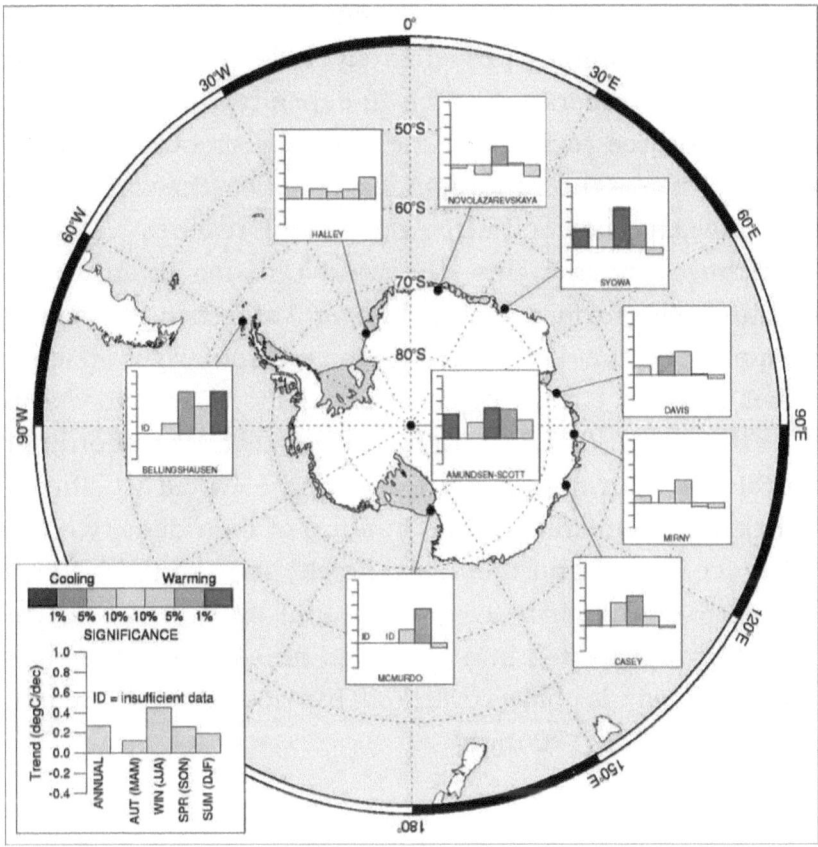

Figure 2.6. Warming at many of the radiosonde stations around the continent

2.3.3. Wind and Pressure

A catabatic wind (named from the Greek word κατάβασις katabasis, meaning "descending") is that the technical name for a drainage wind, a wind that carries high-density air from a better elevation down a slope under the force of gravity. Such winds are sometimes also called fall winds. Katabatic winds can rush down elevated slopes at hurricane speeds, but most aren't as intense as that, and lots of are 10 knots (18 km/h) or less. A catabatic wind originates from radiational cooling of air atop a plateau, a mountain, glacier, or maybe a hill. Since the density of air is inversely proportional to temperature, the air will flow downwards, warming approximately adiabatically because it descends. The temperature of the air depends on the temperature within the source region and therefore the amount of descent. Within the case of the Santa Ana, for instance, the wind can (but doesn't always) become hot by the time it reaches water level. In Antarctica, against this, the wind remains intensely cold. The entire near-surface wind field over Antarctica is essentially determined by the katabatic winds, particularly outside the summer season, except in coastal regions when storms may impose their own wind field. Katabatic winds are most ordinarily found blowing out from the massive and elevated ice sheets of Antarctica and Greenland. The buildup of high density cold air over the ice sheets and therefore the elevation of the ice sheets brings into play enormous gravitational energy. Where these winds are concentrated into restricted areas within the coastal valleys, the winds blow overflow hurricane force, reaching around 300 km/h (190 mph).

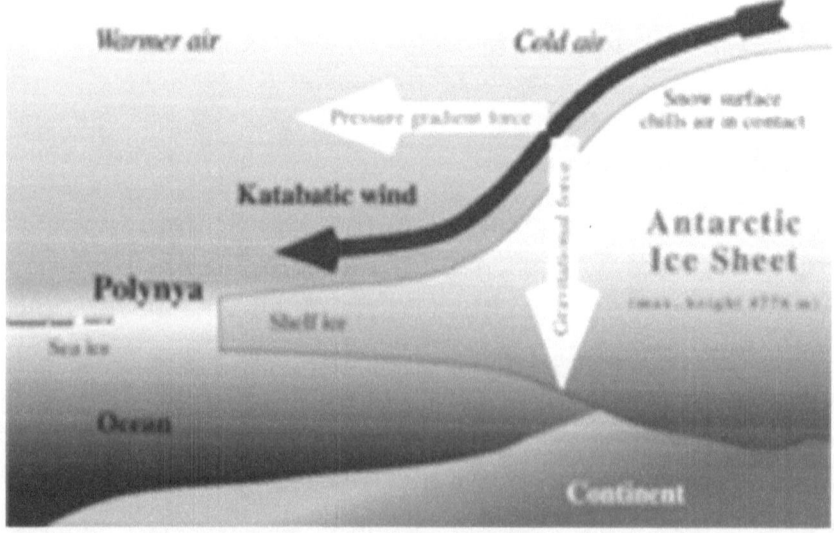

Figure 2.7: Generation of katabatic winds in Antarctica

In a few regions of continental Antarctica the snow is scoured away by the force of the katabatic winds, resulting in "dry valleys" (or "Antarctic oases") like the McMurdo Dry Valleys. Since the katabatic winds are descending, they have a tendency to possess a coffee ratio, which desiccates the region. Other regions may have an identical but lesser effect, resulting in "blue ice" areas where the snow is removed and therefore the surface ice sublimates, but is replenished by glacier be due upstream.

In the Fuegian Archipelago in South America also as in Alaska in North America, a wind referred to as a williwaw may be a particular danger to harboring vessels. Williwaws originate within the snow and ice fields of the coastal mountains, and that they are often faster than 120 knots (140 mph; 220 km/h)

Atmospheric surface pressure provides strong insight into Antarctic climate variability and alters, given its dynamic relationship to the strength and direction of the winds, which influence patterns of thermal advection and sea ice variability. Several studies have indicated changes in air pressure across Antarctica since the late 1970s, including a positive trend within the Southern Annular Mode (SAM) (Marshall, 2003; Jones., 2016), a deepening of the Amundsen Sea Low (Turner, 2009, 2013; Fogt, 2014; Fogt, 2015; Raphael, 2016), and pressure decreases at many coastal East Antarctic stations during austral summer and autumn. Through geostrophic relationships, these pressure changes have caused shifts within the near-surface winds, which are linked to changes in sea ice extent and concentration (Turner 2009; Holland 2012; Meehl 2016), and therefore the resulting changes in thermal advection have similarly been linked to ongoing warming across West Antarctica (Steig et al., 2009; Bromwich et al., 2007) and therefore the Antarctic Peninsula, also because the lack of serious warming across much of East Antarctica (Marshall, 2007; Steig, 2009) since the International Geophysical Year (IGY, 1957/1958).

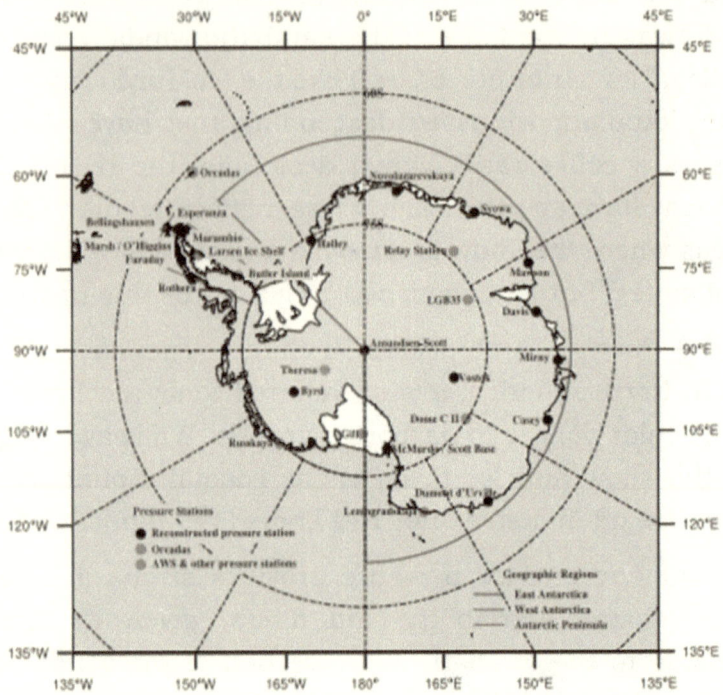

Figure 2.8. Displayed are the reconstructed station pressure locations (green), observations from Orcadas (red), and AWS locations (purple) used to create or evaluate (for AWS) the spatial Antarctic-wide pressure reconstruction. Outlined are the geographic regions used for further comparison, East Antarctica (45°W eastward to 180°, poleward of 66°S), West Antarctica (45°W westward to 180°, poleward of 75°S), and the Antarctic Peninsula (55°W–68°W, 62°S–75°S).

Unfortunately, little or no is understood about the complete spatial pattern of air pressure before 1979, due largely to the very fact that a lot of gridded pressure data sets are unreliable before 1979 over the Antarctic (Bromwich, 2007; Bracegirdle, 2008), including those data sets that reach back throughout the first twentieth century (Allan, 1993; Jones, 2006). to supply a historical context for ongoing pressure changes at the main Antarctic research stations, a recent study reconstructed mean seasonal pressure at 18 stations across Antarctica using midlatitude pressure data as predictors. However, thanks to the massive distances between the stations, these studies didn't provide information about pressure across the whole Antarctica. Here we extend this earlier work by generating a replacement summer pressure reconstruction poleward of 60°S for the amount 1905–2013 using these station reconstructions as anchoring points during a proven interpolation scheme and comprehensively evaluate the summer pressure variability and alter across the whole Antarctica during the complete twentieth century.

A newly generated reliable summer pressure reconstruction across Antarctica since 1905 demonstrates that only across East Antarctica are the recent pressure trends unique when viewed within the context of the complete twentieth century. within the coastal Atlantic sector of West Antarctica, pressures were similarly low within the early twentieth century and, when combined with higher pressure anomalies within the middle twentieth century, produce to significant positive pressure trends in much of the first twentieth century during this region, which are likely caused by natural variability before the amount of Antarctic stratospheric ozone depletion. When examining causality of the Antarctic pressure variability and alter during a suite of climate model experiments within the context of the complete twentieth century, it becomes clear that multiple influences additionally to ozone depletion have had a pronounced impact on Antarctic summer pressure variability, especially within the last 60 years. because the negative summer pressure trends are ubiquitous across Antarctica after 1957 in our reconstruction and observations and project strongly onto a positive SAM pattern, we therefore also suggest that multiple influences additionally to ozone depletion cause the positive trend within the summer SAM index since 1957. Tropical SST variability has likely played a task in negative pressure anomalies across Antarctica within the last 15 years, perhaps timed with the shift of the Pacific Decadal Oscillation/Interdecadal Pacific Oscillation (Meehl et al., 2016) other recent modeling work further highlights the importance of the tropical SSTs compared to ozone depletion after 1979 on the changes within the 850 hPa zonal winds. Our work also suggests that volcanic activity or other radiative forcing mechanisms played a crucial role in generating positive summer pressure anomalies within the late 1960s across Antarctica that contribute to the negative pressure trend within the last 60 years. Future work motivated by the seasonal Antarctic pressure reconstructions as compared with model simulations and estimates of tropical SST variability will hopefully further improve the understanding of 20th century Antarctic climate variability, especially in other seasons outside austral summer where Antarctic atmospheric circulation trends appear more likely regarding natural processes.

2.3.4. Cloud and Precipitation

Clouds are a crucial a part of the worldwide climate system. for instance, the fourth assessment report issued by the Intergovernmental Panel on global climate change (IPCC) noted that the indirect effect of aerosols on clouds may be a relatively large negative radiative forcing of -0.7 Wm-2, but that the range of uncertainty during this value is large and therefore the level of scientific understanding is "low" (IPCC 2007: fig. SPM2). The feedbacks related to clouds, and whether clouds act to warm or cool the atmosphere, are complex, and aren't properly understood even at mid-latitudes where they need been observed intimately for several years. within the Antarctic, cloud observations have largely been confined to synoptic observations, although in recent years these measurements are supplemented by occasional in place microphysical measurements, mainly of low cloud from the surface, and radiometric measurements made up of the surface and from satellite data. These measurements have only been made at a couple of locations—the South Pole and a few coastal stations—and these locations might not be representative of the continent as an entire. The microphysical properties (shape, size, concentration and phase, i.e., whether solid or liquid) of cloud particles can have a serious impact on the Earth's radiation budget, so it's important that they're correctly represented within climate models. However, the parameterizations of clouds utilized in global climate models have normally been developed using measurements made in mid-latitudes, and these might not be applicable to Antarctic clouds. Again, due to the isolated and harsh environment of the continent few in place measurements are of the microphysical properties of clouds within the Antarctic. Polar stratospheric clouds (PSCs) play a central role within the formation of the hole within the Antarctic and Arctic. PSCs provide surfaces upon which heterogeneous chemical reactions happen. These reactions cause the assembly of free radicals of chlorine within the stratosphere which directly destroy ozone molecules.

PSCs form poleward of about 60°S latitude within the altitude range 10 km to 25 km during the winter and early spring. The clouds are classified into Types I and II consistent with their particle size and formation temperature. Type II clouds, also referred to as nacreous or mother-of-pearl clouds, are composed of ice crystals and form when temperatures are below the ice frost point (typically below −83°C). the sort I PSCs are optically much thinner than the sort II clouds, and have a formation threshold temperature 5 to 8°C above the frost point. These clouds consist mainly of hydrated droplets of aqua fortis and vitriol. Despite 20 years of research, the climatology of PSCs isn't well described, and this impacts on the accuracy of ozone depletion models. The timing and duration of PSC events, their geographic extent and vertical distributions, and their annual variability aren't well understood. The Davis LIDAR has been wont to study stratospheric clouds since 2001. The observations contain profiles of Rayleigh laser backscatter at a wavelength of 532 nm as a function of altitude. The measurements are getting used to research the climatology of the clouds and their reference to the temperature structure of the stratosphere, and therefore the influence of atmospheric gravity waves and planetary waves in modulating their structure and ozone depletion. The Australian Antarctic Division encourages people travelling to Antarctica to stay a lookout for these clouds, and to report any sightings. This information is potentially useful in comparing with observations by the Davis LIDAR, satellite measurements and predictions of atmospheric models.

Precipitation

Map of average annual precipitation on Antarctica (mm liquid equivalent) the entire precipitation on Antarctica, averaged over the whole continent, is about 166 millimetres (6.5 inches) per annum. the particular rates vary widely, from high values over the Peninsula (15 to 25 inches a year) to very low values (as little as 50 millimetres (2.0 inches) within the high interior (Bromwich, Reviews of Geophysics, 1988). Areas that receive not up to 250 millimetres (9.8 inches) of precipitation per annum are classified as deserts. most Antarctic precipitation falls as snow. Rainfall is rare and mainly occurs during the summer in coastal areas and surrounding islands. Note that the quoted precipitation may be a measure of its equivalence to water, instead of being the particular depth of snow. The air in Antarctica is additionally very dry. The low temperatures end in a really low absolute humidity, which suggests that dry skin and cracked lips are a continuing problem for scientists and expeditioners working within the continent.

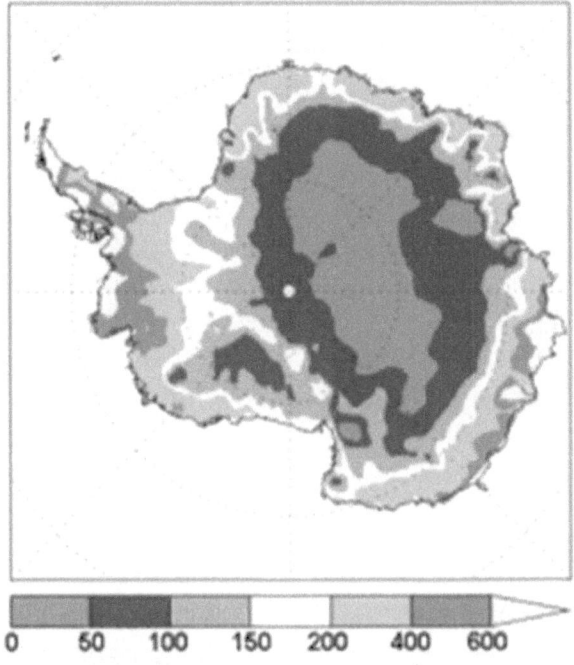

| 0 | 50 | 100 | 150 | 200 | 400 | 600 |

Figure 2.9: Average annual precipitation on Antarctica (mm liquid equivalent)

2.3.5 Ozone Depletion

consists of two related events observed since the late 1970s: a steady lowering of about four percent in the total amount of ozone in Earth's atmosphere (the ozone layer), and a much larger springtime decrease in stratospheric ozone around Earth's polar regions. The latter phenomenon is referred to as the ozone hole. There are also springtime polar tropospheric ozone depletion events in addition to these stratospheric events. In 2019, NASA announced that the "ozone hole" over Antarctica was the smallest ever since it was first discovered in 982The main cause of ozone depletion and the ozone hole is manufactured chemicals, especially manufactured halocarbon refrigerants, solvents, propellants and foam-blowing agents (chlorofluorocarbons (CFCs), HCFCs, halons), and referred to as ozone-depleting substances (ODS). These compounds are transported into the stratosphere by turbulent mixing after being emitted from the surface, mixing much faster than the molecules can settle. Once in the stratosphere, they release halogen atoms through photo dissociation, which catalyze the breakdown of ozone (O3) into oxygen (O2). Both types of ozone depletion were observed to increase as emissions of halocarbons increased.

Ozone depletion and the ozone hole have generated worldwide concern over increased cancer risks and other negative effects. The ozone layer prevents most harmful UV wavelengths of ultraviolet light (UV light) from passing through the Earth's atmosphere. These wavelengths cause skin cancer, sunburn, permanent blindness and cataracts, which were projected to increase dramatically as a result of thinning ozone, as well as harming plants and animals. These concerns led to the adoption of the Montreal Protocol in 1987, which bans the production of CFCs, halons and other ozone-depleting chemicals. The ban came into effect in 1989. Ozone levels stabilized by the mid-1990s and began to recover in the 2000s, as the shifting of the jet stream in the southern hemisphere towards the south pole has stopped and might even be reversing. Recovery is projected to continue over the next century, and the ozone hole is expected to reach pre-1980 levels by around 2075. The Montreal Protocol is considered the most successful international environmental agreement to date.

I) Ozone hole and its causes

The Antarctic ozone hole is an area of the Antarctic stratosphere in which the recent ozone levels have dropped to as low as 33 percent of their pre-1975 values. The ozone hole occurs during the Antarctic spring, from September to early December, as strong westerly winds start to circulate around the continent and create an atmospheric container. Within this polar vortex, over 50 percent of the lower stratospheric ozone is destroyed during the Antarctic spring. As explained above, the primary cause of ozone depletion is the presence of chlorine-containing source gases (primarily CFCs and related halocarbons). In the presence of UV light, these gases dissociate, releasing chlorine atoms, which then go on to catalyze ozone destruction. The Cl-catalyzed ozone depletion can take place in the gas phase, but it is dramatically enhanced in the presence of polar stratospheric clouds (PSCs).]

These polar stratospheric clouds form during winter, in the extreme cold. Polar winters are dark, consisting of three months without solar radiation (sunlight). The lack of sunlight contributes to a decrease in temperature and the polar vortex traps and chills air. Temperatures hover around or below -80 °C. These low temperatures form cloud particles. There are three types of PSC clouds nitric acid trihydrate clouds, slowly cooling water-ice clouds, and rapid cooling water ice (nacreous) clouds provide surfaces for chemical reactions whose products will, in the spring lead to ozone destruction.. The photochemical processes involved are complex but well understood. The key observation is that, ordinarily, most of the chlorine in the stratosphere resides in "reservoir" compounds, primarily chlorine nitrate ($ClONO_2$) as well as stable end products such as HCl. The formation of end products essentially removes Cl from the ozone depletion process. The former sequester Cl, which can be later made available via absorption of light at shorter wavelengths than 400 nm. During the Antarctic winter and spring, however, reactions on the surface of the polar stratospheric cloud particles convert these "reservoir" compounds into reactive free radicals (Cl and ClO). The process by which the clouds remove NO_2 from the stratosphere by converting it to nitric acid in the PSC particles, which then are lost by sedimentation, is called denitrification. This prevents newly formed ClO from being converted back into $ClONO_2$.

The role of sunlight in ozone depletion is the reason why the Antarctic ozone depletion is greatest during spring. During winter, even though PSCs are at their most abundant, there is no light over the pole to drive chemical reactions. During the spring, however, the sun comes out, providing energy to drive photochemical reactions and melt the polar stratospheric clouds, releasing considerable ClO, which drives the whole mechanism. Further warming temperature near the end of spring break the vortex around mid-December. As warm, ozone and NO2-rich air flows in from lower latitudes, the PSCs are destroyed, the enhanced ozone depletion process shuts down, and the ozone hole closes. Most of the ozone that is destroyed is in the lower stratosphere, in contrast to the much smaller ozone depletion through homogeneous gas phase reactions, which occurs primarily in the upper stratosphere.

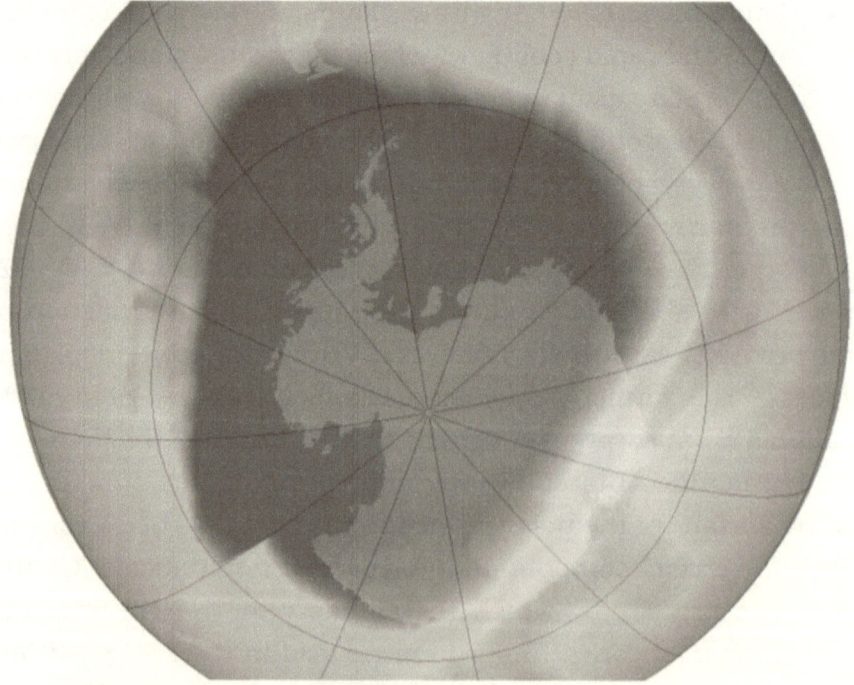

Figure 2.10: The ozone layer in Antarctica (oceanwide-expeditions.com)

The discovery of the Antarctic "ozone hole" by British Antarctic Survey reported during a paper in Nature in May 1985 came as a shock to the scientific community, because the observed decline in polar ozone was far larger than anyone had anticipated. Satellite measurements showing massive depletion of ozone round the South Pole were becoming available at an equivalent time. However, these were initially rejected as unreasonable by data internal control algorithms (they were filtered out as errors since the values were unexpectedly low); the hole was detected only in satellite data when the data was reprocessed following evidence of ozone depletion in in place observations. When the software was rerun without the flags, the hole was seen as far back as 1976. Susan Solomon, an atmospheric chemist at the National Oceanic and Atmospheric Administration (NOAA), proposed that chemical reactions on polar stratospheric clouds (PSCs) within the cold Antarctic stratosphere caused a huge, though localized and seasonal, increase within the amount of chlorine present in active, ozone-destroying forms. The polar stratospheric clouds in Antarctica are only formed when there are very low temperatures, as low as −80 °C, and early spring conditions. In such conditions the ice crystals of the cloud provide an appropriate surface for conversion of unreactive chlorine compounds into reactive chlorine compounds, which may deplete ozone easily.

Since 1981 the United Nations Environment Programme, under the auspices of the planet Meteorological Organization, has sponsored a series of technical reports on the Scientific Assessment of Ozone Depletion, supported satellite measurements. The 2007 report showed that the opening within the ozonosphere was recovering and therefore the smallest it had been for a few decades. The 2010 report found, "Over the past decade, global ozone and ozone within the Arctic and Antarctic regions is not any longer decreasing but isn't yet increasing. The ozonosphere outside the Polar Regions is projected to recover to its pre-1980 levels a while before the center of this century. In contrast, the springtime hole over the Antarctic is predicted to recover much later." In 2012, NOAA and NASA reported "Warmer air temperatures high above the Antarctic led to the second smallest season hole in 20 years averaging 17.9 million square kilometers. The opening reached its maximum size for the season on Sept 22, stretching to 21.2 million square kilometers." A gradual trend toward "healing" was reported in 2016 than in 2017. It's reported that the recovery signal is clear even within the ozone loss saturation altitudes. In 2019, NASA reported that hole shrunk to the littlest size since 1982, and there was no significant relation between its size and global climate change. The opening within the Earth's ozonosphere over the South Pole has affected atmospheric circulation within the hemisphere all the thanks to the equator. The hole has influenced atmospheric circulation all the way to the tropics and increased rainfall at low, subtropical latitudes within the hemisphere.

CHAPTER III

SPACE SCIENCE RESEARCH IN ANTARCTIC

3.1 The Earth's atmosphere

The Earth's atmosphere is a gaseous envelope that surrounds it and acts as a protective shield for the Earth that allows the life to exist on it. At ground level, the atmosphere is composed of Nitrogen (78%), Oxygen (21%), and Argon (1%). Other important components are water (0 - 7%), Ozone (0 - 0.01%) and Carbon Dioxide (0.01-0.1%). Due to the presence of gravity, the atmosphere of the Earth is horizontally stratified. The regions of the neutral atmosphere are categorized by the variations of the composition and the state of mixing and also with the variations of temperature versus height. As one move upwards through the layers, atmospheric pressure and the neutral composition decreases exponentially with altitude. Based on temperature, the atmosphere is divided into four layers: the troposphere, stratosphere, mesosphere, and thermosphere.

1. Troposphere

The troposphere is that the lowest layer of Earth's atmosphere. It extends from surface to a mean height of about 12 km (7.5 mi; 39,000 ft), although this altitude varies from about 9 km (5.6 mi; 30,000 ft) at the geographic poles to 17 km (11 mi; 56,000 ft) at the Equator,[18] with some variation thanks to weather. The troposphere is bounded above by the tropopause, a boundary marked in most places by a temperature inversion (i.e. a layer of relatively warm air above a colder one), and in others by a zone which is isothermal with height.

Although variations do occur, the temperature usually declines with increasing altitude within the troposphere because the troposphere is usually heated through energy transfer from the surface. Thus, rock bottom a part of the troposphere (Earth's surface) is usually the warmest section of the troposphere. The troposphere contains roughly 80% of the mass of Earth's atmosphere. The troposphere is denser than all its overlying atmospheric layers because a bigger atmospheric weight sits on top of the troposphere and causes it to be most severely compressed. one-half of the entire mass of the atmosphere is found within the lower 5.6 km (3.5 mi; 18,000 ft) of the troposphere.

Nearly all atmospheric water vapor or moisture is found within the troposphere, so it's the layer where most of Earth's weather takes place. it's basically all the weather-associated cloud genus types generated by active wind circulation, although very tall cumulonimbus thunder clouds can penetrate the tropopause from below and rise into the lower a part of the stratosphere. Most conventional aviation activity takes place within the troposphere, and it's the sole layer which will be accessed by propeller-driven aircraft

II. Stratosphere

The stratosphere is that the second-lowest layer of Earth's atmosphere. It lies above the troposphere and is separated from it by the tropopause. This layer extends from the highest of the troposphere at roughly 12 km (7.5 mi; 39,000 ft) above surface to the stratopause at an altitude of about 50 to 55 km (31 to 34 mi; 164,000 to 180,000 ft).

The air pressure at the highest of the stratosphere is roughly 1/1000 the pressure stumped level. It contains the ozonosphere, which is that the a part of Earth's atmosphere that contains relatively high concentrations of that gas. The stratosphere defines a layer during which temperatures rise with increasing altitude. This rise in temperature is caused by the absorption of ultraviolet (UV) radiation from the Sun by the ozonosphere, which restricts turbulence and mixing. Although the temperature could also be −60 °C (−76 °F; 210 K) at the tropopause, the highest of the stratosphere is far warmer, and should be near 0 °C.The stratospheric temperature profile creates very stable atmospheric conditions, therefore the stratosphere lacks the weather-producing air turbulence that's so prevalent within the troposphere. Consequently, the stratosphere is nearly completely freed from clouds and other sorts of weather. However, polar stratospheric or nacreous clouds are occasionally seen within the lower a part of this layer of the atmosphere where the air is coldest. The stratosphere is that the highest layer which will be accessed by jet-powered aircraft.

III. Mesosphere

The mesosphere is that the third highest layer of Earth's atmosphere, occupying the region above the stratosphere and below the thermosphere. It extends from the stratopause at an altitude of about 50 km (31 mi; 160,000 ft) to the mesopause at 80–85 km (50–53 mi; 260,000–280,000 ft) above water level. Temperatures drop with increasing altitude to the mesopause that marks the highest of this middle layer of the atmosphere. it's the coldest place on Earth and has a mean temperature around −85 °C (−120 °F; 190 K). Just below the mesopause, the air is so cold that even the very scarce water vapors at this altitude are often sublimated into polar-mesospheric noctilucent clouds. These are the very best clouds within the atmosphere and should be visible to the eye if sunlight reflects off them about an hour or two after sunset or an identical length of your time before sunrise. they're most readily visible when the Sun is around 4 to 16 degrees below the horizon. Lightning-induced discharges referred to as transient luminous events occasionally form within the mesosphere above tropospheric thunderclouds. The mesosphere is additionally the layer where most meteors spend upon atmospheric entrance. It's too high above Earth to be accessible to jet-powered aircraft and balloons, and too low to allow orbital spacecraft. The mesosphere is especially accessed by sounding rockets and rocket-powered aircraft.

IV. Thermosphere

The thermosphere is that the second-highest layer of Earth's atmosphere. It extends from the mesopause (which separates it from the mesosphere) at an altitude of about 80 km (50 mi; 260,000 ft) up to the thermopause at an altitude range of 500–1000 km (310–620 mi; 1,600,000–3,300,000 ft). The peak of the thermopause varies considerably thanks to changes in solar activity.[17] Because the thermopause lies at the lower boundary of the exosphere, it's also mentioned because the exobase. The lower a part of the thermosphere, from 80 to 550 kilometres (50 to 342 mi) above surface, contains the ionosphere. The temperature of the thermosphere gradually increases with height. Unlike the stratosphere beneath it, wherein a temperature inversion is thanks to the absorption of radiation by ozone, the inversion within the thermosphere occurs thanks to the extremely rarity of its molecules. The temperature of this layer can rise as high as 1500 °C (2700 °F), though the gas molecules are thus far apart that its temperature within the usual sense isn't very meaningful. The air is so rarefied that a private molecule (of oxygen, for example) travels a mean of 1 kilometer (0.62 mi; 3300 ft) between collisions with other molecules.[19] Although the thermosphere features a high proportion of molecules with high energy, it might not feel hot to a person's in direct contact, because its density is just too low to conduct a big amount of energy to or from the skin. This layer is totally cloudless and freed from water vapour. However, non-hydro meteorological phenomena like the northern lights and southern lights are occasionally seen within the thermosphere. The International space platform orbits during this layer, between 350 and 420 km (220 and 260 mi).

V. Exosphere

The exosphere is that the outermost layer of Earth's atmosphere (i.e. the upper limit of the atmosphere). It extends from the exobase, which is found at the highest of the thermosphere at an altitude of about 700 km above water level, to about 10,000 km (6,200 mi; 33,000,000 ft) where it merges into the solar radiation. This layer is especially composed of extremely low densities of hydrogen, helium and a number of other heavier molecules including nitrogen, oxygen and CO2 closer to the exobase. The atoms and molecules are thus far apart that they will travel many kilometers without colliding with each other. Thus, the exosphere not behaves sort of a gas, and therefore the particles constantly escape into space. These free-moving particles follow ballistic trajectories and should migrate in and out of the magnetosphere or the solar radiation. The exosphere is found too far above Earth for any meteorological phenomena to be possible. However, the northern lights and southern lights sometimes occur within the lower a part of the exosphere, where they overlap into the thermosphere. The exosphere contains most of the satellites orbiting Earth.

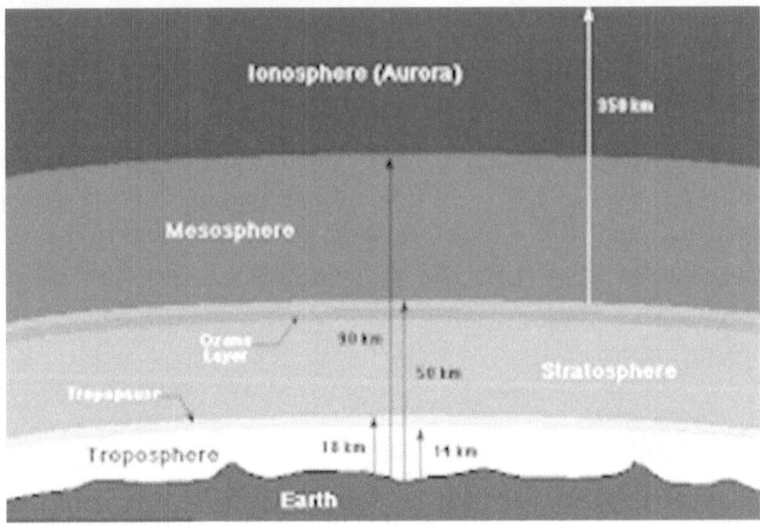

The Earth's Atmosphere

3.2. Ionosphere

The ionosphere is that the ionized a part of Earth's upper atmosphere, from about 60 km to 1,000 km altitude, a neighborhood that has the thermosphere and parts of the mesosphere and exosphere. The ionosphere is ionized by radiation. It plays a crucial role in electrical discharge and forms the inner fringe of the magnetosphere. it's practical importance because, among other functions, it influences radio propagation to foreign places on the world. The region below the ionosphere is named neutral atmosphere, or neutrosphere. Discovery of the ionosphere extended over nearly a century. As early as 1839, the German mathematician Carl Friedrich Gauss speculated that an electrically conducting region of the atmosphere could account for observed variations of Earth's magnetic flux. The notion of a conducting region was reinvoked by others, notably in 1902 by the American engineer Arthur E. Kennelly and therefore the English physicist Heaviside, to elucidate the transmission of radio signals round the curve of Earth's surface before definitive evidence was obtained in 1925. For a few years the ion-rich region was mentioned because the Heaviside layer.

The name "ionosphere" was introduced first within the 1920s and was formally defined in 1950 by a committee of the Institute of Radio Engineers as "the part of the earth's upper atmosphere where ions and electrons are present in quantities sufficient to affect the propagation of radio waves." Much of the first research on the ionosphere was administered by radio engineers and was stimulated by the necessity to define the factors influencing long-range radio communication. Subsequent research has focused on understanding the ionosphere because the environment for Earth-orbiting satellites and, within the military arena, for missile flight. knowledge domain of the ionosphere has grown tremendously, fueled by a gentle stream of knowledge from spacecraft-borne instruments and enhanced by measurements of relevant atomic and molecular processes within the laboratory.

Historically, the ionosphere was thought to be composed of variety of relatively distinct layers that were identified by the letters D, E, and F. The Appleton layer was subsequently divided into regions F1 and F2. It's now known that each one these layers aren't particularly distinct, but the first naming scheme persists. It appears that Edward V. Appleton, a pioneer in early radio probing of the ionosphere, is liable for the nomenclature. Appleton was familiar with using the symbol E to explain the electrical field of the wave reflected from the primary layer of the ionosphere that he studied. Later he identified a second layer at higher altitude and used the symbol F for the reflected wave. Suspecting a layer at lower altitude, he adopted the extra symbol D. In time, the letters came to be related to the layers themselves instead of with the sector of the reflected waves. it's now known that electron density increases more or less uniformly with altitude from the D-layer, reaching a maximum within the F2 region. Though the nomenclature wont to describe the various layers of the ionosphere continues in wide use, the definitions have evolved to reflect the improved understanding of the underlying physics and chemistry.

3.3. D region

The D-layer is that the lowest ionospheric region, at altitudes of about 70 to 90 km (40 to 55 miles). The D-layer differs from the E and F regions therein its free electrons almost totally disappear during the night, because they recombine with oxygen ions to make electrically neutral oxygen molecules. At this point, radio waves undergo to the strongly reflecting E and F layers above. During the day some reflection are often obtained from the D-layer, but the strength of radio waves is reduced; this is often the explanation for the marked reduction within the range of radio transmissions in daytime. At its upper boundary the D-layer merges with the Heaviside layer.

3.4. E region

The Heaviside layer is additionally called Heaviside layer, named for American engineer Arthur E. Kennelly and English physicist Heaviside in 1902. It extends from an altitude of 90 km (60 miles) to about 160 km (100 miles). Unlike that of the D-layer, the ionization of the Heaviside layer remains in the dark, though it's considerably diminished. The Heaviside layer was liable for the reflections involved in Guglielmo Marconi's original transatlantic radio communication in 1902. The ionization density is usually 105 electrons per milliliter during the day, though intermittent patches of stronger ionization are sometimes observed.

3.5. F region

The Appleton layer extends upward from an altitude of about 160 km (100 miles). This region has the best concentration of free electrons. Although its degree of ionization persists with little change through the night, there's a change within the ion distribution. During the day, two layers are often distinguished: a little layer referred to as F1 and above it a more highly ionized dominant layer called F2. in the dark they merge at about the extent of the F2 layer, which is additionally called the Appleton layer. This region reflects radio waves with frequencies up to about 35 megahertz; the precise value depends on the height amount of the electron concentration, typically 106 electrons per milliliter, though with large variations caused by the sunspot cycle.

3.6. Magnetosphere

Earth's magnetosphere began in 1600, when William Gilbert discovered that the magnetic flux on the surface of Earth resembled that on a Terrell, a small, magnetized sphere. Within the 1940s, Walter M. Elsasser proposed the model of dynamo theory, which attributes Earth's magnetic flux to the motion of Earth's iron outer core. Through the utilization of magnetometers, scientists were ready to study the variations in Earth's magnetic flux as functions of both time and latitude and longitude.

Beginning within the late 1940s, rockets were wont to study cosmic rays. In 1958, Explorer 1, the primary of the Explorer series of space missions, was launched to review the intensity of cosmic rays above the atmosphere and measure the fluctuations during this activity. This mission observed the existence of the Van Allen radiation belt (located within the inner region of Earth's magnetosphere), with the follow up Explorer 3 later that year definitively proving its existence. Also during 1958, Eugene Parker proposed the thought of the solar radiation, with the term 'magnetosphere' being proposed by Thomas Gold in 1959 to elucidate how the solar radiation interacted with the Earth's magnetic flux. The later mission of Explorer 12 in 1961 led by the

Cahill and Amazeen observation in 1963 of a sudden decrease in magnetic flux strength near the noon-time meridian later was named the magnetopause. By 1983, the International Cometary Explorer observed the magnetotail, or the distant magnetic flux

The overall structure of the outer ionosphere the magnetosphere is strongly influenced by the configuration of Earth's magnetic flux. On the brink of the planet's surface, the magnetic flux features a structure almost like that of a perfect dipole. Field lines are oriented more or less vertically at high latitudes, sweep back over the Equator, where they're essentially horizontal, and hook up with Earth during a symmetrical pattern at high latitudes. The sector departs from this ideal dipolar configuration, however, at high altitudes. There the terrestrial field, Earth's magnetic flux, is distorted to a big extent by the solar radiation, with its embedded solar magnetic flux. Ultimately the terrestrial field is dominated by the interplanetary field, which is generated by the Sun. Over Earth's equator, the magnetic flux lines become almost horizontal, and then return to reconnect at high latitudes. However, at high altitudes, the magnetic flux is significantly distorted by the solar radiation and its solar magnetic flux. On the dayside of Earth, the magnetic flux is significantly compressed by the solar radiation to a distance of roughly 65,000 kilometers (40,000 mi). Earth's bow shock is about 17 kilometers (11 mi) thick and located about 90,000 kilometers (56,000 mi) from Earth. The magnetopause exists at a distance of several hundred kilometers above surface. Earth's magnetopause has been compared to a sieve because it allows solar radiation particles to enter. Kelvin–Helmholtz instabilities occur when large swirls of plasma follow the sting of the magnetosphere at a special velocity from the magnetosphere, causing the plasma to slide past. This leads to magnetic reconnection, and because the magnetic flux lines break and reconnect, solar radiation particles are ready to enter the magnetosphere. On Earth's nightside, the magnetic flux extends within the magnetotail, which lengthwise exceeds 6,300,000 kilometers (3,900,000 mi). Earth's magnetotail is that the primary source of the polar aurora. Also, NASA scientists have suggested that Earth's magnetotail might cause "dust storms" on the Moon by creating a possible difference between the day side and therefore the night side. The solar radiation compresses the magnetic flux on Earth's dayside at a distance of about 10 Earth radii, or almost 65,000 km from the earth. At this distance the magnetic flux is so weak that the pressure related to particles escaping from Earth's gravity is like the opposing pressure related to the solar radiation. This equilibrium region, with a characteristic thickness of 100 km (60 miles), is named the magnetopause and marks the periphery of the magnetosphere. The lower boundary of the magnetosphere is several hundred kilometres above Earth's surface. On the nightside, the terrestrial field is stretched during a giant tail that reaches past the orbit of the Moon, extending perhaps to distances in more than 1,000 Earth radii. The magnetotail can reach such great distances because on the nightside the forces related to the magnetic flux and therefore the solar radiations are parallel.

The outermost regions of the magnetosphere are exceedingly complex, especially at high latitudes, where terrestrial field lines are hospitable space. Ionization from the solar radiation can leak into the magnetosphere during a number of the way. It can enter by turbulent exchange at the dayside magnetopause or more directly at cusps within the magnetopause at high latitudes where closed loops of the magnetic flux on the dayside meet fields connecting to the magnetotail. Additionally, it can enter at large distances on the nightside, where the magnetic pressure is comparatively low and where field lines can reconnect readily, providing quick access to the enormous plasma sheet within the interior of Earth's magnetotail. The magnetosheath, a neighborhood of magnetic turbulence during which both the magnitude and therefore the direction of Earth's magnetic flux vary erratically, occurs between 10 and 13 Earth radii toward the Sun. This disturbed region is assumed to be caused by the assembly of magneto hydrodynamic shock waves, which successively are caused by high-velocity solar radiation particles. Before this bow shock boundary, toward the Sun, is that the undisturbed solar radiation.

Earth's magnetic field

Earth's magnetic flux, also referred to as the geomagnetic field, is that the magnetic flux that extends from the Earth's interior out into space, where it interacts with the solar radiation, a stream of charged particles emanating from the Sun. The magnetic flux is generated by electric currents thanks to the motion of convection currents of molten iron within the Earth's outer core: these convection currents are caused by heat escaping from the core, a natural action called a geodynamo. The magnitude of the Earth's magnetic flux at its surface ranges from 25 to 65 microteslas (0.25 to 0.65 gauss). As an approximation, it's represented by a field of a dipole currently tilted at an angle of about 11 degrees with reference to Earth's rotational axis, as if there have been a huge magnet placed at that angle through the middle of the world. The North geomagnetic pole, which was in 2015 located on Ellesmere Island, Nunavut, Canada, within the hemisphere, is really the South Pole of the Earth's magnetic flux, and conversely.

While the North and South magnetic poles are usually located near the geographic poles, they slowly and continuously give way geologic time scales, but sufficiently slowly for ordinary compasses to stay useful for navigation. However, at irregular intervals averaging several hundred thousand years, the Earth's field reverses and therefore the North and South Magnetic Poles respectively, abruptly switch places. These reversals of the geomagnetic poles leave a record in rocks that are useful to paleomagnetists in calculating geomagnetic fields within the past. Such information successively is useful in studying the motions of continents and ocean floors within the process of tectonics.The magnetosphere is that the region above the ionosphere that's defined by the extent of the Earth's magnetic flux in space. It extends several tens of thousands of kilometers into space, protecting the world from the charged particles of the solar radiation and cosmic rays that might otherwise strip away the upper atmosphere, including the ozonosphere that protects the world from harmful ultraviolet.

Historically, the north and south poles of a magnet were first defined by the Earth's magnetic flux, not the other way around, since one among the primary uses for a magnet was as a compass needle. A magnet's North Pole is defined because the pole that's attracted by the Earth's North Magnetic Pole when the magnet is suspended so it can turn freely. Since opposite poles attract, the North Magnetic Pole of the world is basically the South Pole of its magnetic flux.

The positions of the magnetic poles are repeatedly defined in a minimum of two ways: nearby or globally. The local definition is that the point where the magnetic flux is vertical. this will be determined by measuring the inclination. The inclination of the Earth's field is 90° (downwards) at the North Magnetic Pole and -90° (upwards) at the South Magnetic Pole. The two poles wander independently of every other and aren't directly opposite one another on the world. Movements of up to 40 kilometers per annum are observed for the North Magnetic Pole. Over the last 180 years, the North Magnetic Pole has been migrating northwestward, from Cape Adelaide within the Boothia Peninsula in 1831 to 600 kilometers from Resolute Bay in 2001.

The magnetic line is that the line where the inclination is zero (the magnetic flux is horizontal).

The global definition of the Earth's field is predicated on a mathematical model. If a line is drawn through the middle of the world, parallel to the instant of the best-fitting dipole, the 2 positions where it intersects the surface are called the North and South geomagnetic poles. If the Earth's magnetic flux were perfectly dipolar, the geomagnetic poles and dip poles would coincide and compasses would point towards them. However, the Earth's field features a significant non-dipolar contribution, therefore the poles don't coincide and compasses don't generally point at either. While the poles are constantly shifting, they need also completely reversed a minimum of a couple of hundred times within the last 3 billion years, consistent with NASA. During this process, which usually occurs every 200,000 to 300,000 years over the course of 100 to a couple of thousand years at a time, the magnetic flux becomes squashed and pulled with multiple poles sprouting up randomly over the surface of the world. The last full reversal occurred about 780,000 years ago. The history of the magnetic flux, including shifts and reversals, is evidenced within the geologic record. Metals found in rocks, including iron, align with the magnetic flux before molten rocks solidify or as fragments that contain the magnetic metals aligned with the magnetic flux and settle in layers of sedimentary rocks. "Since the world may be a dynamic and ever-changing place, new rocks, and their magnetic records, are generated constantly throughout geological time," Smirnov said, adding that these records are often preserved for millions or billions of years. Similar records are found on the ground of the Atlantic where new seafloor is consistently being created at the mid-Atlantic ridge. "As the lava wells up to the surface [through the long crack that creates up the ridge], it's molten, and therefore the iron particles suspended within the lava orient themselves within the direction of Earth's prevailing magnetic flux," Ingram said. Because the lava solidifies, it locks the metal deposits in situ, and thus, creates a historic record of the shifts and reversals of Earth's magnetic flux.

3.7. Auroras

Auroras are the results of disturbances within the magnetosphere caused by solar radiation. These disturbances are sometimes strong enough to change the trajectories of charged particles in both solar radiation and magnetospheric plasma. These particles, mainly electrons and protons, precipitate into the upper atmosphere (thermosphere/exosphere).The resulting ionization and excitation of atmospheric constituents emit light of varying color and complexity. the shape of the aurora, occurring within bands around both polar regions, is additionally hooked in to the quantity of acceleration imparted to the precipitating particles. Precipitating protons generally produce optical emissions as incident hydrogen atoms after gaining electrons from the atmosphere. Proton auroras are usually observed at lower latitudes (Sakaguchi, 2007).

Most auroras occur during a band referred to as the "auroral zone", which is usually 3° to 6° wide in latitude and between 10° and 20° from the geomagnetic poles in the least local times (or longitudes), most clearly seen in the dark against a dark sky. A neighborhood that currently displays an aurora is named the "auroral oval", a band displaced towards the night side of the world. Early evidence for a geomagnetic connection comes from the statistics of auroral observations., established that the aurora appeared mainly within the auroral zone. Day-to-day positions of the auroral ovals are posted on the web.

Aurora over Antarctica

In northern latitudes, the effect is understood because the northern lights or the aurora borealis. the previous term was coined by Galileo in 1619, from the Roman goddess of the dawn and therefore the Greek name for the norther (Siscoe, 1986). The southern counterpart, the southern lights or the aurora australis, has features almost just like the northern lights and changes simultaneously with changes within the northern auroral zone (Østgaard, 2007). The southern lights are visible from high southern latitudes in Antarctica, Chile, Argentina, New Zealand, and Australia.

A geomagnetic storm causes the auroral ovals (north and south) to expand, bringing the aurora to lower latitudes. The instantaneous distribution of auroras (Feldstein, (2011)) is slightly different, being centered about 3–5° nightward of the magnetic pole, in order that auroral arcs reach furthest toward the equator when the magnetic pole in question is in between the observer and therefore the Sun. The aurora are often seen best at this point, which is named magnetic midnight.

Auroras seen within the auroral oval could also be directly overhead, but from farther away, they illuminate the poleward horizon as a greenish glow, or sometimes a faint red, as if the Sun were rising from an unusual direction. Auroras also occur poleward of the auroral zone as either diffuse patches or arcs, which may be subvisual.

Auroras are occasionally seen in latitudes below the auroral zone, when a geomagnetic storm temporarily enlarges the auroral oval. Large geomagnetic storms are commonest during the height of the 11-year sunspot cycle or during the three years after the height.(Stamper, 2000, Papitashvili,1999)

An electron spirals (gyrates) a few line of force at an angle that's determined by its velocity vectors, parallel and perpendicular, respectively, to the local geomagnetic field vector B. This angle is understood because the "pitch angle" of the particle. the space, or radius, of the electron from the sector line at any time is understood as its Larmor radius. The pitch angle increases because the electron travels to a neighborhood of greater field intensity nearer to the atmosphere. Thus, it's possible for a few particles to return, or mirror, if the angle becomes 90° before entering the atmosphere to hit the denser molecules there. Other particles that don't mirror enter the atmosphere and contribute to the auroral display over a variety of altitudes. Other sorts of auroras are observed from space, "poleward arcs" stretching sunward across the polar cap, the related "theta aurora" and "dayside arcs" near noon. These are relatively infrequent and poorly understood. Other interesting effects occur like flickering aurora, "black aurora" and subvisual red arcs. Additionally to all or any these, a weak glow (often deep red) observed round the two polar cusps, the sector lines separating those that close through the world from people who are swept into the tail and shut remotely.

A full understanding of the physical processes which cause differing types of auroras remains incomplete, but the essential cause involves the interaction of the solar radiation with the Earth's magnetosphere. The varying intensity of the solar radiation produces effects of various magnitudes but includes one or more of the subsequent physical scenarios.

A quiescent solar radiation flowing past the Earth's magnetosphere steadily interacts with it and may both inject solar radiation particles directly onto the geomagnetic field lines that are 'open', as against being 'closed' within the opposite hemisphere, and supply diffusion through the bow shock. It also can cause particles already trapped within the radiation belts to precipitate into the atmosphere. Once particles are lost to the atmosphere from the radiation belts, under quiet conditions, new ones replace them only slowly, and therefore the loss-cone becomes depleted. Within the magnetotail, however, particle trajectories seem constantly to reshuffle, probably when the particles cross the very weak magnetic flux near the equator. As a result, the flow of electrons therein region is almost an equivalent altogether direction and assures a gentle supply of leaking electrons. The leakage of electrons doesn't leave the tail charged, because each leaked electron lost to the atmosphere is replaced by a coffee energy electron drawn upward from the ionosphere. Such replacement of "hot" electrons by "cold" ones is in complete accord with the 2nd law of thermodynamics. the entire process, which also generates an electrical ring current round the Earth, is uncertain.

Geomagnetic disturbance from an enhanced solar radiation causes distortions of the magnetotail. These 'substorms' tend to occur after prolonged spells (hours) during which the interplanetary magnetic flux has had an appreciable southward component. This results in a better rate of interconnection between its field lines and people of Earth. As a result, the solar radiation moves magnetic flux (tubes of magnetic flux lines, 'locked' alongside their resident plasma) from the day side of Earth to the magnetotail, widening the obstacle it presents to the solar radiation flow and constricting the tail on the night-side. Ultimately some tail plasma can separate "magnetic reconnection" some plasmoids are squeezed downstream and are over excited with the solar wind; others are squeezed toward Earth where their motion feeds strong outbursts of auroras, mainly around midnight. A geomagnetic storm resulting from greater interaction adds more particles to the plasma trapped around Earth, also producing enhancement of the "ring current". Occasionally the resulting modification of the Earth's magnetic flux are often so strong that it produces auroras visible at middle latitudes, on field lines much closer to the equator than those of the auroral zone.

Acceleration of auroral charged particles invariably accompanies a magnetospheric disturbance that causes an aurora. This mechanism, which is believed to predominantly arise from strong electric fields along the magnetic flux or wave-particle interactions, raises the speed of a particle within the direction of the guiding magnetic flux. The pitch angle is thereby decreased and increases the prospect of it being precipitated into the atmosphere. Both electromagnetic and electrostatic waves, produced at the time of greater geomagnetic disturbances, make a big contribution to the energizing processes that sustain an aurora. Particle acceleration provides a posh intermediate process for transferring energy from the solar radiation indirectly into the atmosphere

The details of those phenomena aren't fully understood. However, it's clear that the prime source of auroral particles is that the solar radiation feeding the magnetosphere, the reservoir containing the radiation zones and temporarily magnetically-trapped particles confined by the geomagnetic field, including particle acceleration processes.

The electrons liable for the brightest sorts of the aurora are well accounted for by their acceleration within the dynamic electric fields of plasma turbulence encountered during precipitation from the magnetosphere into the auroral atmosphere. In contrast, static electric fields are unable to transfer energy to the electrons thanks to their conservative nature.(Bryant, 1998). The electrons and ions that cause the diffuse aurora appear to not be accelerated during precipitation. The rise in strength of magnetic flux lines towards the world creates a 'magnetic mirror' that turns back many of the downward flowing electrons. The brilliant sorts of auroras are produced when downward acceleration not only increases the energy of precipitating electrons but also reduces their pitch angles (angle between electron velocity and therefore the local magnetic flux vector). This greatly increases the speed of deposition of energy into the atmosphere, and thereby the rates of ionization, excitation and consequent emission of auroral light. Acceleration also increases the electron current flowing between the atmosphere and magnetosphere.

One early theory proposed for the acceleration of auroral electrons is predicated on an assumed static, or quasi-static, field creating a uni-directional electric potential. No mention is provided of either the required space-charge or equipotential distribution, and these remain to be specified for the notion of acceleration by double layers to be credible. Fundamentally, Poisson's equation indicates that there are often no configurations of charge leading to a net electric potential. Inexplicably though, some authors still invoke quasi-static parallel electric fields as net accelerators of auroral electrons, citing interpretations of transient observations of fields and particles as supporting this theory as firm fact. In another example,[60] there's little justification given for saying 'FAST observations demonstrate detailed quantitative agreement between the measured electric potentials and therefore the ionic beam energies, leaving little question that parallel potential drops are a dominant source of auroral particle acceleration'.

Another theory is predicated on acceleration by Landau, resonance within the turbulent electric fields of the acceleration region. This process is actually an equivalent as that employed in plasma fusion laboratories throughout the planet,[62] and appears well ready to account in theory for many – if not all – detailed properties of the electrons liable for the brightest sorts of auroras, above, below and within the acceleration region

Other mechanisms have also been proposed, especially, Alfvén waves, wave modes involving the magnetic flux first noted by Hannes Alfvén (1942), which are observed within the laboratory and in space. The question is whether or not these waves might just be a special way of watching the above process, however, because this approach doesn't means a special energy source, and lots of plasma bulk phenomena also can be described in terms of Alfvén waves. Other processes also are involved within the aurora, and far remain to be learned. Auroral electrons created by large geomagnetic storms often seem to possess energies below 1 keV and are stopped above, near 200 km. Such low energies excite mainly the line of oxygen in order that often such auroras are red. On the opposite hand, positive ions also reach the ionosphere at such time, with energies of 20–30 keV, suggesting they could be an "overflow" along magnetic flux lines of the copious "ring current" ions accelerated at such times, by processes different from those described above. Some O+ ions ("conics") also seem accelerated in several ways by plasma processes related to the aurora. These ions are accelerated by plasma waves in directions mainly perpendicular to the sector lines. They, therefore, start at their "mirror points" and may travel only upward. As they are doing so, the "mirror effect" transforms their directions of motion, from perpendicular to the sector line to a cone around it, which gradually narrows down, becoming increasingly parallel at large distances where the sector is far weaker.

3.8. Meteor Shower

Meteorite in antarctica

Scientists estimate that about 48.5 tons (44 tonnes or 44,000 kilograms) of meteoritic material falls on the world every day. Most the fabric is vaporized in Earth's atmosphere, leaving a bright trail fondly called "shooting stars." Several meteors per hour can usually be seen on any given night. Sometimes the amount increases dramatically—these events are termed meteor showers. Meteor shows occur annually or at regular intervals because the Earth passes through the trail of dusty debris left by a comet. Meteor showers are usually named after a star or constellation that's on the brink of where the meteors appear within the sky. Perhaps the foremost famous are the Perseids, which peak in August per annum. Every Perseid meteor may be a tiny piece of the comet Swift-Tuttle, which swings by the Sun every 135 years.The first Antarctic meteorite was discovered during the 1911-14 Douglas Mawson Australasian Antarctic Expedition. Subsequent discoveries of three more were made in 1961 and

1964 by Russian geologists near Novolazarevskaya Station, and USGS geologists within the Thiel and Neptune Mountains. In 1969, Japanese explorers discovered nine meteorites at the Yamato Mountains, and 12 more during the 1973-74 season, 663 during the 1974-75 season, and 307 during the 1975-76 season. supported that success, William A. Cassidy received funding for ANSMET to commence within the 1976-77 season. Cassidy was the PI for the 1976-77 seasons, and subsequent seasons up to and including the 1993-94 season. ANSMET (Antarctic look for Meteorites) may be a program funded by the Office of Polar Programs of the National Science Foundation that appears for meteorites within the Transantarctic Mountains. This geographic area is a set point for meteorites that have originally fallen on the extensive high-altitude ice fields throughout Antarctica. Such meteorites are quickly covered by subsequent snowfall and start a centuries-long journey traveling "downhill" across the Antarctica while embedded during a vast sheet of flowing ice. Portions of such flowing ice are often halted by natural barriers like the Transantarctic Mountains. Subsequent wind erosion of the motionless ice brings trapped meteorites back to the surface another time where they'll be collected. This process concentrates meteorites during a few specific areas to much higher concentrations than they're normally found everywhere else. The contrast of the dark meteorites against the white snow, and lack of terrestrial rocks on the ice, makes such meteorites relatively easy to seek out. However, the overwhelming majority of such ice-embedded meteorites eventually slide undiscovered into the ocean.

CHAPTER IV

SPACE WEATHER EFFECTS ON ANTARCTICA

S PACE weather is influenced by the solar radiation and therefore the interplanetary magnetic flux (IMF) carried by the solar radiation plasma. a spread of physical phenomena are related to space weather, including geomagnetic storms and substorms, energization of the Van Allen radiation belts, ionospheric disturbances and scintillation of satellite-to-ground radio signals and long-range radar signals, aurora, and geomagnetically induced currents at surface. Coronal mass ejections (CMEs), their associated shock waves and coronal clouds also are important drivers of space weather as they will compress the magnetosphere and trigger geomagnetic storms. Solar energetic particles (SEP) accelerated by coronal mass ejections or solar flares can trigger solar particle events (SPEs), a critical driver of human impact space weather as they will damage electronics onboard spacecraft (e.g. Galaxy 15 failure), and threaten the lives of astronauts also as increase radiation hazards to high-altitude, high-latitude aviation.

4.1 Space Weather and Sun

Space weather is the concept of changing environmental conditions in near-Earth space. It is distinct from the concept of weather within a planetary atmosphere, and deals with phenomena involving ambient plasma, magnetic fields, radiation and other matter in space. "Space weather" often implicitly means the conditions in near-Earth space within the magnetosphere and ionosphere, but it is also studied in interplanetary (and occasionally interstellar) space. Within our own solar system, space weather is greatly influenced by the speed and density of the solar wind and the interplanetary magnetic field (IMF) carried by the solar wind plasma. A variety of physical phenomena are associated with space weather, including geomagnetic storms and sub storms, energization of the Van Allen radiation belts, ionospheric disturbances and scintillation, aurora and geomagnetically induced currents at Earth's surface. Coronal Mass Ejections and their associated shock waves are also important drivers of space weather as they can compress the magnetosphere and trigger geomagnetic storms. Solar energetic particles, accelerated by coronal mass ejections or solar flares, are also an important driver of space weather as they can damage electronics onboard spacecraft through induced electric currents and threaten the life of astronauts. Space weather exerts a profound influence in several areas related to space exploration and development. Changing geomagnetic conditions can induce changes in atmospheric density causing the rapid degradation of spacecraft altitude in low Earth orbit. Geomagnetic storms due to increased solar activity can potentially blind sensors onboard spacecraft, or interfere with on-board electronics. An understanding of space environmental conditions is also important in designing shielding and life support systems for manned spacecraft. There is also some concern that geomagnetic storms may also expose conventional aircraft flying at high altitudes to increased amounts of radiation.

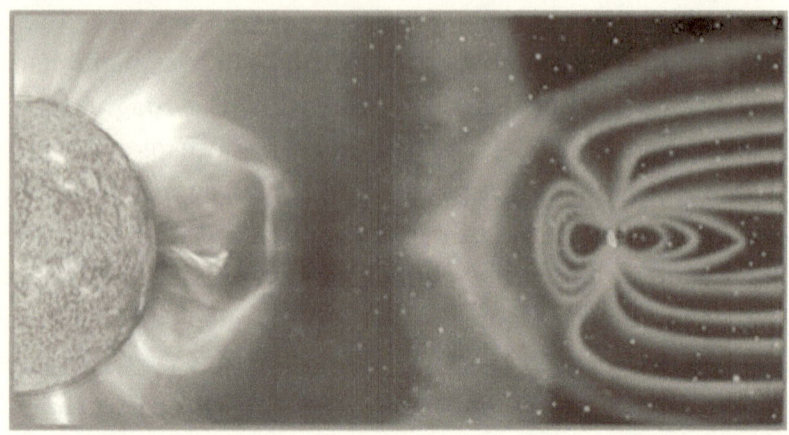

Sun, interplanetary medium and near-Earth environment represents the region in which space weather effected (NASA)

4.1.1 Solar Flares

The high magnetic fields within the sunspot-producing active regions also produce to explosions referred to as solar flares. When the twisted field lines cross and reconnect, energy explodes outward with a force exceeding that of many hydrogen bombs. Temperatures within the outer layer of the sun, referred to as the corona, typically fall around a couple of million kelvins. As solar flares erupt the corona, they heat its gas to anywhere from 10 to twenty million K, occasionally reaching as high as 100 million K. consistent with NASA, the energy released during a flare is the equivalent of many 100-megaton hydrogen bombs exploding at an equivalent time. Because solar flares form within the same active regions as sunspots, they're connected to those smaller, less violent events. Flares tend to follow an equivalent 11-year cycle. At the height of the cycle, several flares may occur every day, with a mean lifetime of only 10 minutes. The big sunspot of 2014 fired off several powerful solar flares.

The largest, X-class flares have the foremost significant effect on Earth. They will cause long-lasting radiation storms within the upper atmosphere, and trigger radio blackouts. Medium-size

M-class flares can cause brief radio blackouts within the polar regions and therefore the occasional minor radiation storms. C-class flares have few noticeable consequences.

When the energized particles exploding from solar flare race toward us, they arrive in just eight minute. Astronauts in space risk being hit by these hazardous particles, and manned missions to the moon or Mars must take this danger under consideration. Everyone else is shielded by the Earth's atmosphere and magnetic flux. Sensitive equipment in space also can be damaged by these energetic particles.

Absorbing X-rays affects the atmosphere. The heat rise and energy end in an expansion of the Earth's ionosphere. Man-made radio waves travel through this portion of the upper atmosphere, so radio communications are often disturbed by its sudden unpredictable growth. Similarly, satellites previously circling through vacuum-free space can find themselves caught within the expanded sphere. The resulting friction slows down their orbit, and may bring them back to Earth before intended. Despite their size and high energy, solar flares are almost never visible optically. The brilliant emission of the encompassing photosphere, where the sun's light originates, tends to overshadow even these explosive phenomena. Radio and optical emissions are often observed on Earth. However, since X-rays and gamma rays fail to penetrate the atmosphere, only space-based telescopes can detect their signatures.

Sometimes, it isn't activity but a scarcity of it which will release deadly particles toward Earth. The interactions of hot plasma of the corona with the sun's magnetic flux can create coronal holes, which enable plasma to stream rapidly from the sun. "The effects linked to coronal holes are generally milder than those of coronal mass ejections, but, when the outflow of solar particles is intense, can pose risk to satellites in orbit," NASA said during a statement. In 2017, scientists were ready to link high-energy gamma radiation bursts to solar flares for the primary time using NASA's Fermi Gamma-ray Space Telescope and its Solar Terrestrial Relations Observatory (STEREO).

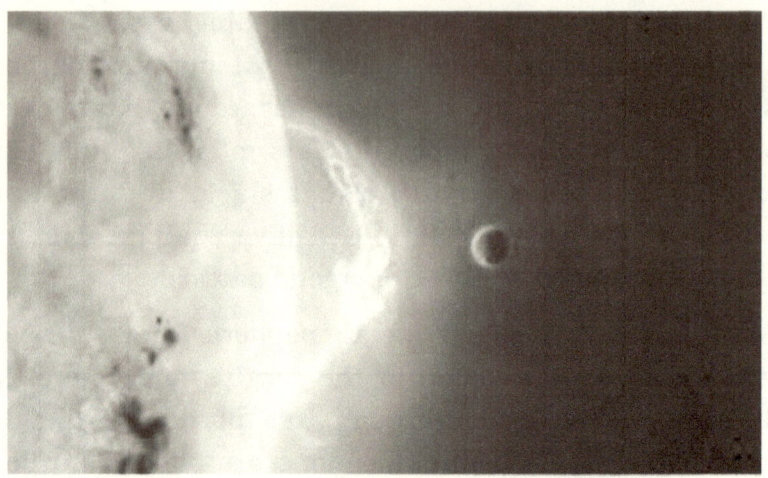

Solar flear Ejecting form the sun

4.1.2 Coronal Mass Ejections

Coronal Mass Ejections (CMEs) are dynamic, large-scale events in the solar corona that expel vast amount of magnetic flux and solar plasma ($\sim 10^{15}$–10^{16} g) into interplanetary space. CMEs typically appear as loop-like features that disrupt helmet streamers in the solar corona. They were first observed with space-based coronagraph in the early 1970's on Orbiting Solar Observatory (OSO) and Skylab. Subsequent observations from the solar wind and Solar Maximum Mission spacecrafts allowed identification of many of the properties of CMEs. The Solar and Heliospheric Observatory (SOHO) spacecraft has now extensively observed CME events from solar minimum in 1996 into the present maximum phase of the solar cycle. Halo events observed with the Large Angle Spectrometric Coronagraphs (LASCO) now provide the most effective means of identifying earthward directed CMEs, which are believed to be primary cause of large, non-recurrent geomagnetic storms. Figure below is a time sequence of SMM coronagraph images showing a typical CME initiation and eruption observed in white light. This CME originates from a helmet streamer that has been slowly rising or swelling outward days before the eruption.

A clear three-part structure of the CME is seen: A bright frontal loop, a dark cavity, and a bright core, which is associated with the prominence that has also erupted. Although not all CMEs originate from where there is a streamer, it is generally believed and observed that CMEs originate from where the field was initially closed but subsequently forced open during the eruption. CMEs have a wide range of measured speeds, ranging from 10 km s^{-1} to over 2000 kms^{-1} with a median speed around 450 kms^{-1} whereas in some extreme events the speeds may be as high as 2700 kms^{-1}. Recently much attention has been given to the question of whether there are two dynamical types of CMEs. One group, the majority in number, is called the "slow CMEs", characterized by relatively lower speeds and observable accelerations in the coronagraph field of view. The other group, a minority in number, is called the "fast CMEs", characterized by relatively higher speeds and insignificant accelerations. This observational classification was first proposed by Mac Queen and Fisher from the Skylab data, but was confirmed by Sheeley (1999) using the LASCO/SOHO data that have a better sensitivity and a larger field of view.

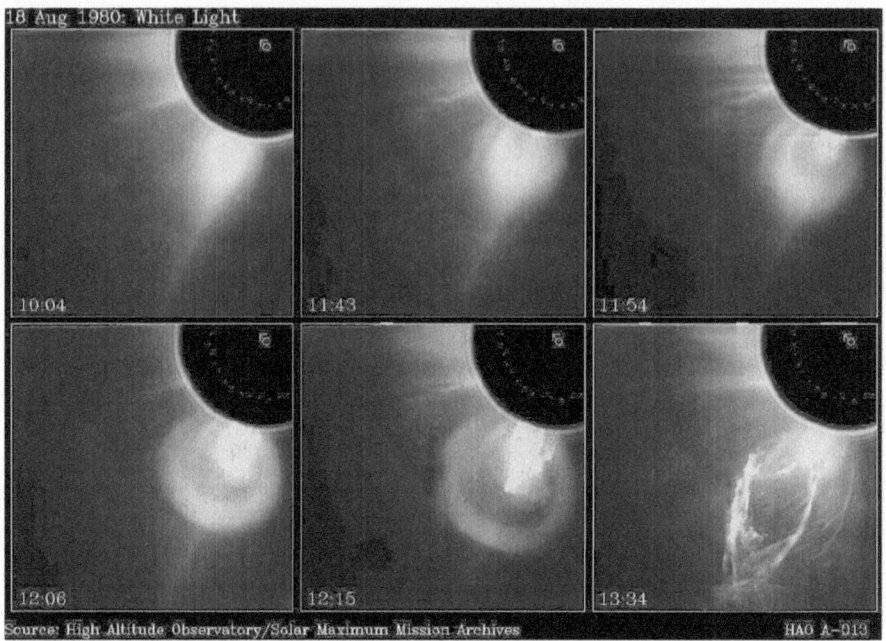

A time sequence of solar maximum emission (SMM-NASA)
coronagraph image showing a coronal mass ejection
on August 18, 1980.

The occurrence of CMEs shows strong solar cycle dependence. There are fewer CMEs occurring during solar minimum, as few as one every two or three days on average. CMEs are more frequently observed during the solar maximum, with several to ten CMEs on some unusual days during the peak period. Averaged over a solar cycle, these observed CMEs are occurring at a rate of two to three events per day. It is worth noting that this rate is lower than the average rate of the occurrence of GOES X-ray flares, which is five to six events per day, averaged over a solar cycle down to GOES B-class flares. Another interesting property of the CMEs is their masses. Unlike the wide range of observed CME speeds, the masses of CMEs are found to lie in a relatively narrow range around 10^{15}–10^{16} g per CME. It is also an interesting question why this should be the case, or why the CME masses do not show something like a power-law distribution, a distribution that is common to many astrophysical quantities. At an average frequency of two or three events per day, the corona is losing its mass through CMEs at an average rate of 10^{11} gs^{-1} or less. Compared to a mean solar-wind mass loss rate of about 2×10^{12} gs^{-1}, the total mass loss through CMEs is not significant, although each CME itself is a significant disturbance to its local environment and solar wind. CMEs also have a strong association with solar flares and erupting prominences. This suggests that these three coronal phenomena may be connected by a common underlying physics. CMEs are distinct in two ways as compared to other mass ejections like the Hα surges, flare sprays, spicules etc.

i)   Each CME is far more massive than any of these mass ejections and the mass in a CME almost always leaves the Sun.

ii) A CME liberates energy of the order of 10^{31}–10^{32} erg, in the form of the work done to lift its mass against gravity and to produce the kinetic energy of the expelled mass.

This energy is comparable to that of a flare, putting the CME with the flare as the two most energetic phenomena in the solar corona. A point to be noted here is that flares and CMEs liberate energies in two distinct forms. In the case of a flare, much of its energy manifests itself as thermalized energy the net result in this case is the intense heating of the corona whereas in case of CME, the liberated energy is in the ordered form of macroscopic work and bulk kinetic energy. In recent years, CMEs have been one of the major topics of solar physics research. Both space-borne and ground-based observations provide a great deal of important data for the study of CME source regions, their initiation, and their early acceleration and propagation. Furthermore, many researchers have used numerical and analytical approaches to investigate the initiation and evolution of CMEs. However, many fundamental questions regarding the nature of CMEs are still unanswered. In particular, the mechanism of CME initiation remains poorly understood.

4.1.3. Flare-CME Relationship

Early in the study of CMEs it was found and later studies confirmed it, that a great many CMEs occur in association with prominence eruptions and flares in their vicinities. The association with eruptive prominences is not surprising since many of the CMEs originate from the disruption of preexisting helmet streamer with a prominence at its base. In a statistical study sing microwave observations have shown that 72 % of all eruptive prominences are associated with CMEs. The association with flares raised few interesting questions. A simple and attractive scenario was proposed in the late seventies that a CME is the part of the corona expelled out by the energy of the associated flare. Early thinking on the solar origin of geomagnetic storms led to the suggestion that a flare might send off a blast wave into interplanetary space to disrupt the Earth's magnetosphere upon arrival. When CMEs were discovered, it was natural to suggest that the CME might be the blast wave in the above scenario. However, observations does not support this suggested identification. Those CMEs moving at extremely low speeds certainly do not fit the description of a blast wave. Careful analysis of the fast moving typical CMEs show that they also do not have the properties of blast waves. In a few cases where it is possible to identify an MHD wave front in the white-light observations, it is found that the wave front is quite distinct from the CME. This is quite consistent with the fact that the CME in most cases is a pre-existing structure, which has broken away as proposed to a dynamically and impulsively created wave front. Moreover, CMEs with or without an associated flare do not look significantly different in white light, suggesting that CMEs and flares are distinct processes. As to the hydro magnetic nature of CMEs, there is a longstanding debate in the solar physics community, which is whether the CMEs are driven by ideal MHD processes or resistive MHD processes. Another way of presenting this question is centered on the flare and CME relationship, that is, whether the flares are the cause of the CMEs or the CMEs are the cause of the flares. By a statistical analysis of the temporal relationships between flares and CMEs, Harrison (1995) concluded that "the flare and CME are the signatures of the same magnetic disease, that is, they represent the responses in different parts of the magnetic structure, to a particular activity: they do not drive one another but are closely related". Recently this question has returned to the attention of solar scientists, partly because space missions like SOHO and TRACE have provided data with better sensitivities, higher temporal and spatial resolutions, to put ideas to further observational tests, and partly because of new studies of physical processes that have benefited from increased computational power. According to Zhang and Low (2005), the question has passed beyond the cause and effect argument. Flares and CMEs are independent MHD processes as Harrison (1995) has pointed out, though they have a great tendency to occur together. So it is more interesting to ask how they occur together and how magnetic reconnection influences the dynamics of CMEs.

4.2 Geomagnetic indices

Geomagnetic indices are designed to describe variation in the geomagnetic field caused by these irregular current systems. Daily regular magnetic field variations arise from current systems caused by regular solar radiation changes. Other irregular current systems produce magnetic field changes caused by

+ The interaction of the solar wind with the magnetosphere,
+ By the magnetosphere itself,

+ By the interactions between the magnetosphere and ionosphere,
+ By the ionosphere itself.

Therefore, magnetic activity indices were designed to describe variation in the geomagnetic field caused by these irregular current systems. Let us give a brief description of other geomagnetic indices which are interesting for the solar-terrestrial relations.

4.2.1. Dst Index

DST stands for Disturbance Storm Time. The DST is an index of magnetic activity derived from a network of near-equatorial geomagnetic observatories that measures the intensity of the globally symmetrical equatorial electrojet (the "ring current"). Thus DST monitors the variations of the globally symmetrical ring current, which encircles the Earth close to the magnetic equator in the Van Allen (or radiation) belt of the magnetosphere. During large magnetic storms the signature of the ring current can be seen in ground magnetic field recordings worldwide as so-called main phase depression. The ring current energization which results in typical depression of 100 nT is related to magnetic reconnection processes at the neutral sheet.

The following plot contains data for the 2003 Halloween super storm with a range of activity levels indicated by the shaded regions. Different researchers may define alternative categories, but the typical cut-off for studies concerned with the effects of geomagnetic storms on power grids and satellites is Dst less than -200 to -300 nT, regardless of the terminology. Compare the quiet-time behaviour on the left side of the plot with the super storm on the right, and keep these ranges in mind when viewing the current Dst data.

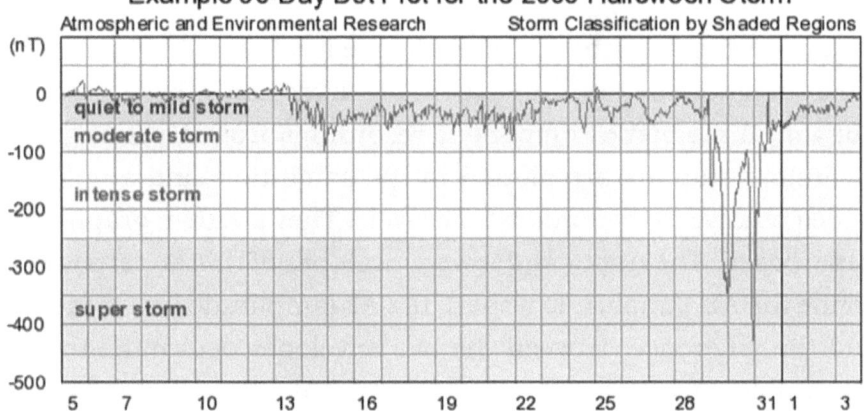

Example 30-Day Dst Plot for the 2003 Halloween Storm

4.2.2. Kp and Ap Index

The K-Index was first introduced by J. Bartels in 1938. It is a quasi-logarithmic local index of the 3-hourly range in magnetic activity relative to an assumed quietday curve for a single geomagnetic observatory site. The values consist of a single digit 0...9 for each 3-hour interval of the universal time day (UT). The planetary 3-hour-range index Kp is the mean standardized K-index from 13 geomagnetic observatories between 44 degrees and 60 degrees northern or southern geomagnetic latitude. The scale is 0...9 expressed in thirds of a unit, e.g. 5- is 4 2/3, 5 is 5 and 5+ is 5 1/3. This planetary index is designed to measure solar particle radiation by its magnetic effects. The 3-hourly Ap (equivalent range) index is derived from the Kp index (see Table 1.5). This table is made in such a way that at a station at about magnetic latitude 50 degrees, Ap may be regarded as the range of the most disturbed of the three field components, expressed in the unit of 2 g. A daily index Ap is obtained by averaging the eight values of Ap for each day. Another index devised to express geomagnetic activity on the basis of the Cp index is the C9 index which has the range between 0 and 9.

4.2.3. AE and other indices

These indices describe the disturbance level recorded by auroral zone magnetometers. In order to determine these indices, horizontal magnetic component recordings from a set of globe-encircling stations are plotted to the same time and amplitude scales relative to their quiet-time levels. They are then graphically superposed. The upper and lower envelopes of this superposition define the AU (amplitude upper), the AL (amplitude lower) indices and the difference between the two envelopes determine the AE (Auroral Electrojet) index, i.e., AE = AU - AL. AO is defined as the average value of AU and AL.

4.3 Solar Indices

The sun emits radio energy with slowly varying intensity. The radio flux, which is originates from atmospheric layers, high in the sun's chromosphere and low in its corona, changes gradually from day to day in response to the number of spot groups on the disk. Solar flux from the entire solar disk at a frequency of 2800 MHz has been recorded routinely by a radio telescope near Ottawa since 10–22Js–1Hz–1 February 1947. The observed values have to be adjusted for the changing Sun-Earth distance and for uncertainties in antenna gain (absolute values).

4.3.1. 10.7 cm Radio Flux

The 10.7 cm solar radio flux, which originates high in the chromosphere of the Sun and low in the corona of the solar atmosphere, is an excellent indicator of solar activity. The F10.7 radio emissions measurements are completely objective, because they can be made in all types of weather. It is one of the longest running records, provides solar activity over six solar cycles, and has proven very valuable in forecasting space weather. There exists a relationship between the solar 10.7 cm radio flux (F 10.7 index) and sunspots. We found a new formula (there is currently none), to calculate the sunspot number from the radio flux.

4.3.2. Sunspot Number

The number of sunspots changes with an 11 years period. Today we know that all solar activity phenomena are related to sunspots and thus to magnetic activity. To measure the solar activity the sunspot numbers were introduced:

$$R = k \, (10g + f)$$

Here g denotes the number of sunspot groups and f the number of spots. The factor is a correction which takes into account for the different instruments used for the determination of R. In order to smear out effects of solar rotation, R is given as a monthly averaged number and called the sunspot relative number. Today there exist better methods to quantify the solar activity however sunspot numbers are available for nearly 400 years and thus this number is still used.

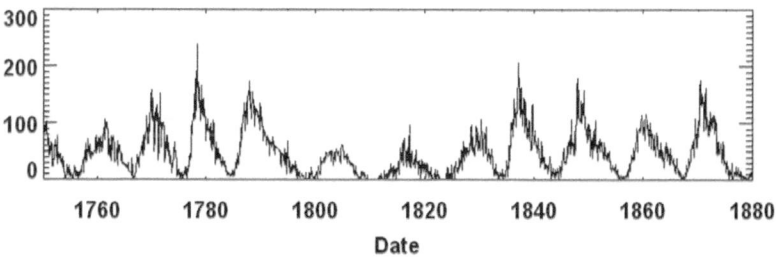

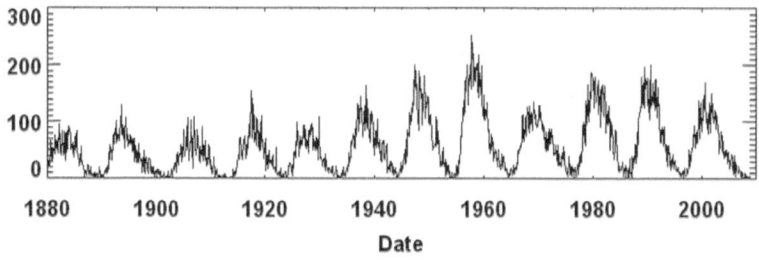

International sunspot numbers from 1745 to the present

CHAPTER V

OBSERVATION AND INSTRUMENTATION

5.1 Satellite Based 63

Understanding what's causing the variations within the atmosphere is extremely important within the technological era we sleep in today. As a society, we are very hooked in to communication and navigation networks round the globe – both space based, and ground based. We've also recently developed a robust reliance on broadcast navigation signals like those provided by GPS satellites. The radio signals used for communication and navigation must propagate through the ionosphere, and non-uniform distributions of plasma within the ionosphere can act like bubbles during a lens or scratches during a mirror, distorting the signal, sometimes to the purpose of unintelligibility

5.1.1 Global Positioning System

GPS is a Global Positioning System based on satellite technology. GPS was developed to replace the TRANSIT system because large time gap in coverage and relatively low navigation accuracy were the main problems with the TRANSIT. The fundamental technique of the GPS is to measure the range between the receiver and a few simultaneously observed satellites. The technique has been very widely applied in several areas such as air, sea and land navigation, Low Earth Orbit Satellites (LEOS) orbit determination, static and kinematics positioning, flight state monitoring as well as surveying etc.

The original objective of the GPS was the instantaneous determination of the position and velocity and the precise coordinate of time. A detailed definition given by W. Wooden in 1985: "The Navstar Global positioning System (GPS) is an all-weather space based navigation system under the development by the Department of Defence (DoD) to satisfy the requirements for military forces to accurately determine their position, velocity and time in common reference system anywhere on or near the earth on a continuous basis". Since DoD is the initiator of GPS, the primary goals were military ones, but the US congress with the guidance from the President, directed DoD to promote its civil use and then GPS has became a necessity for daily life, industries, research and education. GPS satellites broadcast their signals on two carrier frequencies. Simultaneous measurements of the pesudorange and carrier phase of these two signals differ mainly because of the presence of free electrons in the ionosphere.

GPS Signal Frequencies
fo = 10.23 MHz Fundamental frequency
L1 = 1575.42 MHz Primary carrier frequency
L2 = 12276.60 MHz Secondary carrier frequency

All GPS satellites transmit two microwave carrier signals. The signals are broadcast using spread spectrum technique which allows many signals to coexist on the same frequency, and for receivers to detect and separate the different signals from each other.

The L1 primary carrier frequency (1575.42 MHz) carriers the navigation message and the Standard Positioning Services (SPS) code signals. The secondary signals frequency, L2 (12276.60 MHz) is used to measure the Ionospheric delay.

These dual frequency signals, L1 and L2 carriers waves generated by the fundamental frequency of 10.23 MHz by 154 and 120 respectively. The pseduranges that are derived from measured time interval time of the signal from each satellite to the receiver uses two-psedurandom noise codes that are superimposed onto the two carriers.

The standard Coarse/ Acquisition (C/A) GPS code which is available for civilian use also called as civilian code. The C/A code presently modulated upon L1 carrier phase only. The C/A pattern is generated by the hardware signal generator consisting of a pair of 10 bit shift register with feedback connection in them, whose output are combined by an XOR gate. The resulting digital sequence is referred as a Psedu- random Number (PRN). There is a different C/A code or PRN for each satellite, thus the GPS satellites are often identified by their PRN numbers, the unique identifier for each psedurandom noise code.

The second code is Precision code (P code) which has been reserved for U.S. military and other authorized users. The P code modulates both L1 and L2 carrier signals frequencies. The P code designated as the Precise positioning Service (PPS), has an effective wavelength of approximately 30m. This second carrier is 90° ahead of the carrier with the C/A code, but it is of a lower amplitude. In addition P-code sequence is much larger than the C/A code it doesn't repeat over a complete week.

The navigation message also modulates the L1- C/A code signals. The navigation message is a 50 Hz signal consisting of data bits that describe the GPS satellite orbits, clock corrections and other system parameters.

One of the main problem with the TRANSIT system was the fact that it was not able to provide continuous positioning. It has been found that at least four satellites should be always electronically visible for continuous monitoring. Nominal GPS operational constellation consist of 24 satellites that orbit the earth in 12 hours with an altitude of about 20200 km above the earth. The satellites orbits repeat almost the same ground track once each day. The orbit altitude is such that the satellites repeat the same track and configuration over any point approximately 24 hours (4 minutes earlier than each day). There are six orbital planes equally spaced 60° apart and inclined at about 55° with respect to the equatorial plane as shown in Figure. Such constellation provides the user with between five and eight GPS satellites visible from any point on the earth.

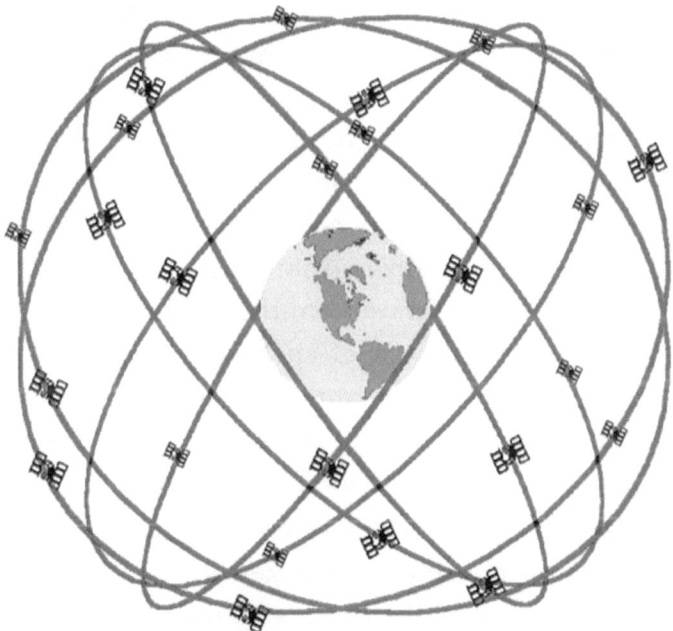

GPS Nominal Constellation
24 Satellites in 6 Orbital Planes
4 Satellites in each Plane
20,200 km Altitudes, 55 Degree Inclination

Figure.5.1. Constellation of GPS satellites

All GPS consist of three elements:

a. The space segment consisting of satellites which broadcast signals.

b. The control segment steering the whole system.

    c. The user segment including different types of GPS receivers.

a. SPACE SEGMENT

The space segments of the system consist of the GPS satellites. These space vehicles send radio signal from space.

The space segment is composed of 24 satellites orbiting the earth twice every sidereal day. These satellites are arranged into 6 high orbital planes at a height of 20200 km. Typically 8-12 satellites visible at one time from anywhere on the earth without any obstruction. The GPS satellites, essentially provide a platform for radio transceivers, atomic clocks, computers, and other various equipment used to operate the system. Each satellite contains a Rubidium atomic clock so between them they represent an extremely accurate time standard available for synchronization at any point on the earth. It is accurate timing that leads to an application of the GPS satellites separate from their function of navigation. The electronic equipment of each satellite allows the user to measure a psedurange to the satellite, and each satellite broadcast a message which allows the user to determine the spatial position of the satellite for arbitrary instants.

Each satellite also transmits a spread spectrum signal containing a Bi-phase switched keyed signal in which 1's & 0's are represented by the reversal of the phase of the carrier. This message is transmitted at the L1 frequency and repeats every 30 minutes called as C/A signal. This message contains two important elements, the almanac and the ephemeris. The almanac gives the information about all the satellites in the constellation and which is regularly updated from ground station monitoring. The ephemeris contains the short lived information about the constellation and the particular satellite sending it and this information is updated from the ground station in every four hour. Besides these there are encrypted signal the P code and Y code that are used for military application transmitted at same frequencies.The satellites have various system of identification such as launch sequence number, assigned pseudorandom noise, orbital position number, NASA catalogue number, and international designation.

b. **CONTROL SEGMENT**

The control segment monitors the signal from the satellites and transmits modifications to the Almanac and Ephemeris as small changes occur in the orbits and the nature of the ionosphere etc. It also monitors the satellite clocks and transmits corrections for these and other parameters necessary to maintain the accuracy of the system. The operational control system consists of a master control station, monitor stations and ground control stations. The main operational tasks for operational control system are:

➢ Tracking of the satellites for the orbit and clock determination and prediction

➢ Time synchronization of the satellites

➢ Upload of the data message to the satellites.

The three most important parts of GPS control segments are

In the beginning the master control station was located at Vandenberg AFB, California, now it has been moved to Consolidated Space Operations Center (CSOC) at Shriver AFB, Colorado Springs, Colorado. The main aim of the CSOC is to collects the tracking data from the monitor stations and then calculates the satellites orbits and clock parameters. These calculated results are then passed to one of the three ground control stations for eventual upload to the satellites. The master control station also controls the satellites and the system operations.

There are five GPS monitor stations located at:

➤ Hawaii

➤ Colorado Springs

➤ Ascension Island in the South Atlantic Ocean

➤ Diego Garcia in the Indian Ocean

➤ Kwajalein in the North Pacific Ocean

Each these stations are equipped with the precise atomic time standard and receivers which continuously measures psedurange to all satellites in view.

These stations are collocated with the monitor stations at Ascension, Diego Garcia, and Kwajalein are the communication links to the satellites and mainly consist of the ground antennas. The satellite ephemerids and clock information calculated from master control station are uploaded to each satellite initially, such uploading was performed every eight hours but now rate has been reduced to once or twice per day.

c. **USERS SEGMENT**

The term user segment is related to the DoD concept of the GPS as an adjunct to the National defense programme. The information transmitted by the ephemeris and almanac a GPS receiver can determine just how long it took the transmitted signals to reach it. This time duration is proportional to the distance of the signal traveled from the satellite so it can be further used to determine an arc on which the receiver must lie. The intersection point of a number of such arcs from different satellites located at different positions provides a solution to the receiver's position on the surface of the earth. The distances calculated for the ranges of the satellites from the raw variables data extracted from the GPS satellites transmissions are called as pseduranges, because the receiver clock's inaccuracy introduces a significant error.

The civilian use of the GPS occurred several years of schedule in manner not envisioned by the system's planners. The primary concept of using an interferometric rather than

Doppler solution model meant that GPS could be used for not only long line geodetic measurements but also for the most exacting short line land survey measurements. Today, GPS receivers are routinely being used to conduct all types of land and geodetic control surveys, and to precisely position photo-aircraft to reduce the amount of ground control needed for mapping.

5.2 Space Based

5.2.1 Geostationary Operational Environmental Satellite (GOES)

The Geostationary Operational Environmental Satellite (GOES), operated by the United States' National Oceanic and Atmospheric Administration (NOAA)'s National Environmental Satellite, Data, and knowledge Service division, supports meteorology, severe storm tracking, and meteorology research. Spacecraft and groundbased element of the system, work together to supply endless stream of environmental data. The National Weather Service (NWS) and therefore the Meteorological Service of Canada use the GOES system for his or her North American weather monitoring and forecasting operations, and scientific researchers use the info to raised understand land, atmosphere, ocean, and climate interactions. The GOES system uses geosynchronous satellites that, since the launch of SMS-1 in 1974, are a basic element of U.S. weather monitoring and forecasting.The procurement, design, and manufacture of GOES satellites is overseen by NASA.NOAA is that the official provider of both GOES terrestrial data and GOES space weather data. Data also can be accessed using the SPEDAS software.

Figure 5.2: Geostationary Operational Environmental Satellite

Designed to work in geosynchronous orbit 35,790 kilometres (22,240 mi) above the world, the GOES spacecraft continuously view the continental us, the Pacific and Atlantic Oceans, Central America, South America, and southern Canada. The three-axis, body-stabilized design enables the sensors to "stare" at the world and thus more frequently image clouds, monitor the surface temperature and water vapor fields, and sound the atmosphere for its vertical thermal and vapor structures. The evolutions of atmospheric phenomena are often followed, ensuring real-time coverage of meteorological events like severe local storms and tropical cyclones. The importance of this capability was proven during hurricanes Hugo (1989) and Andrew (1992)

The GOES spacecraft also enhance operational services and improve support for atmospheric science research, numerical weather prediction models, and environmental sensor design and development

5.2.2 Polar Operational Environmental Satellites (POES)

The Polar-orbiting Operational Environmental Satellite (POES) was a constellation of polar orbiting weather satellites funded by the National Oceanic and Atmospheric Administration (NOAA) and therefore the European Organization for the Exploitation of Meteorological Satellites (EUMETSAT) with the intent of improving the accuracy and detail of weather analysis and forecasting. The Spacecraft were provided by NASA and therefore the European Space Agency, and NASA's Goddard Space Flight Center oversaw the manufacture, integration and test of the NASA-provided TIROS satellites. the primary polar-orbiting meteorological satellite launched as a part of the POES constellation was the tv Infrared Observation Satellite (TIROS), which was launched on April 1, 1960. the ultimate spacecraft, NOAA-19, was launched in February 2009. The ESA-provided MetOp satellite operated by EUMETSAT utilizes POES-heritage instruments for the aim of knowledge continuity. The Joint Polar Satellite System (JPSS), which was launched on November 18, 2017, is that the successor to the POES Program.

On-orbit satellite operations of POES is performed by NOAA's Office of Satellite and merchandise Operations (OSPO)

Figure 5.3. Polar Operational Environmental Satellites (POES)

Data from the POES support a broad range of environmental monitoring applications including weather analysis and forecasting, climate research and prediction, global sea surface temperature measurements, atmospheric soundings of temperature and humidity, ocean dynamics research, eruption monitoring, fire detection, global vegetation analysis, search and rescue, and lots of other applications.

One of the key instruments of the present POES MetOp-B system is that the High Resolution infrared Sounder (HIRS/4). HIRS/4 senses within 20 channels starting from visible bands to radio wave infrared (0.69-14.96 micron wavelengths), to sense variation of temperature, humidity, and pressures within the atmosphere. The info collected from HIRS/4 is collaboratively used with the Advanced Microwave Sounding Instrument (AMSU) to advance research in sea surface temperatures, cloud coverage analysis, ozone concentrations throughout the atmosphere and earth's radiance.

5.3 Ground Based

The concept of ionospheric sounding was born as early as 1924, when Briet and Tuve (1926) proved the existence of an ionized layer with the reception of ionospheric echoes of HF pulses transmitted at 4.3MHz from remote transmitter (distance 13.8km). Thus for more than four decades, sounding the ionosphere with the ionospheric sounders or ionosonde RADAR, DIGISONDE has been the most important technique developed for the investigation of the global structure of ionosphere, its diurnal, seasonal and solar cycle changes, and its response to solar disturbances.

5.3.1 RADAR

Radar may be a detection system that uses radio waves to work out the range, angle, or velocity of objects. It are often wont to detect aircraft, ships, spacecraft, guided missiles, automobiles, weather formations, and terrain. A radar system consists of a transmitter producing electromagnetic waves within the radio or microwaves domain, a transmitting antenna, a receiving antenna (often an equivalent antenna is employed for transmitting and receiving) and a receiver and processor to work out properties of the object(s). Radio waves (pulsed or continuous) from the transmitter reflect off the thing and return to the receiver, giving information about the object's location and speed.

Radar was developed secretly for military use by several nations within the period before and through war II. A key development was the cavity magnetron within the uk, which allowed the creation of relatively small systems with sub-meter resolution. The term RADAR was coined in 1940 by the us Navy as an acronym for RAdio Detection And Ranging.The term radar has since entered English and other languages as a standard noun, losing all capitalization.The following derivation was also suggested during RAF RADAR courses in 1954/5: at Yatesbury Training Camp: Radio Azimuth Direction And Ranging. the fashionable uses of radar are highly diverse, including air and terrestrial control, radar astronomy, air-defense systems, antimissile systems, marine radars to locate landmarks and other ships, aircraft anti-collision systems, ocean surveillance systems, space surveillance and rendezvous systems, meteorological precipitation monitoring, altimetry and control systems, missile target locating systems, and ground-penetrating radar for geological observations. High tech radar systems are related to digital signal processing, machine learning and are capable of extracting useful information from very high noise levels. Radar may be a key technology that the self-driving systems are mainly designed to use, alongside sonar and other sensors.

Other systems almost like radar make use of other parts of the spectrum. One example is LIDAR, which uses predominantly infrared from lasers instead of radio waves. With the emergence of driverless vehicles, radar is predicted to help the automated platform to watch its environment, thus preventing unwanted incidents

The Mesosphere-Stratosphere-Troposphere (MST) radar may be a high power coherent pulse Doppler radar capable of mapping the structure, vector wind fields and turbulence within the atmosphere with very high temporal and spatial resolution. The MST radar contains a two-dimensional phased antenna array, a group of high power transmitters with appropriate feed network, T/R switches, a phase coherent receiver with quadrature channels, a sign processor consisting of two identical channels of AID converter, decoder and integrator, a computer interface and a computer with essential peripherals and software support. MST radar provides estimates of atmospheric winds on endless basis with high temporal and spatial resolution, which is vital within the study of the various dynamical processes of the atmosphere. MST radar uses the echoes obtained over the altitude ranges of 1-100 km to review winds, waves, turbulence and

atmospheric stability. Echoes below 50 km arise primarily thanks to neutral turbulence whereas above 50 km, the echoes are thanks to irregularities within the electron density. In the height ranges 30-60 km, density of the atmosphere also as electron density, are very low and therefore the echoes are very weak, leading to a niche region in most of the MST radars. For probing this region, MST radar alongside Rawinsonde, Dropsonde, Rocketsonde, Lidar and Meteor radar might be used. Woodman and Guillen (1974) studied the lower atmosphere using the incoherent

scatter radar, which is employed to probe the ionosphere. they might obtain echoes from the variation within the index of refraction of the clear air. The contribution of MST radars in the study of the structure and dynamics of middle atmosphere was reviewed by Rottger (1980). MST Radars are often utilized for observing wind, waves and turbulence (Gage and Balsley, 1978). Balsley and Garello (1986) analysed the short period wind fluctuations over poker Flat, Alaska using the Poker Flat MST Radar. The vertical velocity power spectra was studied by Ecklund. (1986) using poker flat MST Radar. 30 Using Indian MST Radar big variety of observations were administered during past few years by variety of scientists. a number of the important topics are study of gravity waves and tidal waves, tropopause detection, study of unstable layers, convection events and ionospheric irregularities. during this work Indian MST Radar was operated in ST mode to review the speed profiles and wave activity.

MST Radar in Atmospheric Studies

The MST radar technique are often considered as having evolved from the pioneering work of Woodman and Guillen (1974). Since then, the technique has been employed by variety of observers to deduce a spread of important properties like wind, waves turbulence and stability of the atmosphere over increasingly greater height ranges. Results obtained from such observations are helpful during a number of disciplines including meteorology, atmospheric dynamics global circulation, gravitation wave and turbulent studies. Gage and Balsley (1978) have discussed the historical perspective of technique, while Balsley and Gage (1980), Harper and Gorden (1980), and Balsley (1981) have considered the potential of the technique for middle atmospheric studies. Related wind measurement techniques are utilized by Gregory (1979), Walker (1979), Harper and Gorden (1980) and Gage and Vanzandt (1981).

Using Indian MST radar at Gadanki within the ST mode of operation, wind velocity measurements were administered by many workers. Study of momentum flux and turbulence parameters were investigated by Jivrajani, (1994). Narayana Rao, (1994) studied the refractivity turbulence structure constant and turbulent energy dissipation rate with Indian MST radar. Characteristics of unstable modes around a shear layer were studied by Mini (1994). The Indian MST radar is additionally used for probing the ionospheric irregularities. Viswanathan (1994) and Rao (1994 b) studied plasma irregularities within the Heaviside layer for the primary time using simultaneous day time observation made during February March 1994 by the MST radar and VHF back scatter radar at Trivandrum. Dynamics of the equatorial spread-F was observed by Patra (1994) using MST radar in ionospheric coherent back scatter mode. A weakening of tropopause and associated enhancement in troposphere stratosphere exchange was observed on some nights and explained as thanks to enhanced turbulence caused by strong wind shears (Jaya Rao, 1995). This phenomenon is reported to be most conspicuous under enhanced convection (Jain,1997). Using the radar wind data, preliminary studies are made from the varied aspects of the lower and middle atmospheric dynamics, including gravity waves, tides and equatorial waves (Narayana Rao,1997; Jivarajani,1997; Sasi,1997). Narayana Rao (1997) have derived eddy dissipation rates using the radar data collected on 17 June 1994. The values are found to vary with height within the range of $10^{\cdot}<>$ to 10.3 m ^{2} s^3 with maximum occurring within the height range of 13 to 16 km. the likelihood that temperature profile are often derived from MST radar data of vertical winds has been acknowledged by Rottger (1986) and demonstrated by Revathy (1996), using the info taken at Gadanki. The derived temperature profile was found to be in good agreement with the radiosonde observations. 49 The mesospheric echo characteristics are studied by Dutta, (1997) and winds and turbulence by Sasi and Vijayan (1997). the foremost intense echoes were generally, confined to a band of 70 - 75 km. The echo characteristics suggested that they were of turbulent scattering instead of of Fresnel reflection. The radar back scatter from meteor trails has been studied by Raghava reddy and Muraleedharan Nair (1998) using the Indian MST radar. a reasonably sizable amount of meteor echoes are detected over the observational windows. The MST radar at Gadanki has been operated in ionospheric coherent back scatter mode for mapping the structure and dynamics of the Heaviside layer and Field aligned irregularities (F AI). it had been shown by Krishna Murthy (1998) that the observed drift velocities below 95 km are driven by neutral wind and therefore the meridional wind component derived from the drift velocity is found to be according to the theoretical neutral wind models. The sector aligned Appleton layer irregularities were studied by Patra (1997) and Rao (1997).

5.3.2 IONOSONDE

An ionosonde, may be a special radar for the examination of the ionosphere. the essential ionosonde technology was invented in 1925 by Gregory Breit and Merle A. Tuve [1] and further developed within the late 1920s by variety of prominent physicists, including Edward Victor Appleton. The term ionosphere and hence, the etymology of its derivatives, was proposed by Robert Watson-Watt..for quite four decades, sounding the ionosphere with the ionospheric sounders or ionosonde has been the foremost important technique developed for the investigation of the worldwide structure of ionosphere, its diurnal, seasonal and solar cycle changes, and its response to solar disturbances. at the present there are various radio techniques, which are wont to study the various properties of ionosphere, but the radio sounding of ionosphere using ionosonde is that the commonest one used throughout the planet for its continuous monitoring. Perhaps, no other ground based experiment has made such substantial contribution to know the ionosphere as ionosonde and remains remain to be so especially with the supply of present day computer controlled advanced digital ionosonde with automatic scaling capability. On contrary, modern techniques of complex ionospheric parameters measurements and processing (Bibl and Reinisch, 1978a; Wright and Pitteway, 1979) have cause resurgence of interest in ionospheric sounding as a basic research tool, while a renewed interest in HF communication is resulting in a rejuvenation of the worldwide ionosonde network.

Basically a sounder may be a sort of radar, which is capable of obtaining echoes from the ionosphere over a good range of operating frequencies. Ionosonde determines the ionospheric vertical profile from the bottom up to the peak of maximum electron density. Its working is predicated on the reflection of radio emission by the ionospheric conducting plasma. The reflection happens when plasma frequency is bigger than the radio sounding frequency, since the plasma frequency is that the function of electron density hence these reflections are often wont to sound the ionospheric electron density profile.

With an ionosonde, a brief pulse of radio emission is transmitted vertically upward and received at an equivalent place after reflection from the ionosphere. The frequency is modified steadily over a couple of minutes and therefore the time period of the heart beat is photographically recorded because the ionogram. Ionogram, may be a record of ionospheric condition indicated by the connection between the radio pulse emitted upward and therefore the virtual height of echoes reflected from the ionosphere. Ionograms are often wont to determine the electron density distribution as a function of height, from a height that's approximately rock bottom of the Heaviside layer to generally the height of the F2 layer. in additional convenient, ionosonds are often wont to determine the propagation conditions on HF communication links. In an ionogram the horizontal scale show the frequency of emitted pulse and therefore the vertical scale represent the effective height (virtual height) at which a sharply reflecting layer would need to be if it were reflect the heart beat with the observed time of interval. The time delays are usually measured in terms of virtual distance by the relation h=ct/2, c being the speed of electromagnetic waves during a vacuum. The number h would be adequate to the range of reflection if the propagation occurred at speed c, which isn't true. Thus the peak scale of ionogram is marked on the idea that the radio waves propagate within the ionosphere at a speed of sunshine. As a matter of fact, however, they propagate more slowly in such an ionized medium because the ionosphere. Therefore the recorded height always tends to be above the important height of reflection. Thus the recorded height refers to virtual height (h`), which are indicated as h`F1, h`F2, h`E etc layer by layer. Thus the virtual height of ionospheric layers could also be defined because the height to which a brief pulse of energy sent vertically upward and traveling with the speed of sunshine would reach taking an equivalent two ways time period, as does the particular pulse reflected from the layer, virtual height is usually greater than the important height of ionospheric layers. If the virtual height of the layer is understood, then it's easy to calculate the angle of incidence required for the wave to return to the world at a desired point.

Since ionospheric density varies with time, ionospheric sounding is employed to get information on change in critical frequency and other parameters of the electron density as height profile. The critical frequency is that the limiting frequency below which a radio emission is reflected by, and above which it penetrates and passes through, the ionized medium (an ionospheric layer) at vertical incidence, and it's different for various layer. it's usually denoted by the fo and fc, and for a specific regular layer it's proportional to the root of the utmost electron density of that layer.

The highest frequency which will be reflected back by the ionosphere is one that the index of refraction μ, becomes zero. Of course, the critical frequency is that the highest frequency which may be reflected by a specific layer at vertical incidence but it's, not the very best frequency which can get reflected from the other angle of incidence. Thus the critical gives a thought that radio waves of frequency adequate to or not up to the critical frequency will definitely be reflected back by the ionospheric layer regardless of the angle of incidence. The ordinary and therefore the extra- ordinary critical frequencies of F2 layer are the foremost important and widely studied parameters of the ionosphere. Waves of frequencies exceeding the critical frequency of extra-ordinary Appleton layer (fxf2) can't be reflected at vertical incidence but penetrate the ionosphere completely.

5.3.3 DIGISONDE

The capabilities of recent ionospheric sounders are illustrated using results obtained with the Digisonde 256 and therefore the Digisonde Portable Sounder. Unlike the other ionosonde, these instruments routinely perform a spread of complementary operations in real time: automatic ionogram scaling, calculation of electron density profiles, high resolution Doppler and angle of arrival, plasma drift velocity, polarisation, and precision group height.Today's advanced digital sounders1 are digital HF radars with real time analysis capabilities. This paper shows results from selected stations of the Digisonde network, including the Digisonde 256 (D256) and therefore the Digisonde Portable Sounder (DPS). Since the D256 has previously been described it suffices to present the most differences between the DPS and therefore the D256 (Table 1). Low weight (40 kg), low power (500W peak) and pulse-to-pulse software control of the transmission waveforms make the DPS very attractive for routine monitoring and research.The Digisondes measure all observables of the received signals at each of 128 or 256 ranges: Doppler spectrum (amplitude and phase), angles of arrival, and polarizations. The quiet daytime ionogram from El Arenosillo, Spain, illustrates the format applied for routine sounding: 100 kHz frequency steps, 128 x 5 km height increments. Small optically coded numbers represent the echo amplitudes: X polarisation in grey, O in black. The ARTIST-scaled3 traces are marked by the letters E and F. Superimposed on the ionogram is that the electron density profile. for every range, the DPS measures a full complex spectrum with 2N Doppler lines (N = 1, 2...7) for both O and X polarisation. This information is compressed into a typical ionogram, and for routine archiving each echo amplitude is amended by Doppler, polarization and angle of incidence. The Ny Alesund ionogram, strong spread on the "overhead" echoes also as an off-vertical patch of ionization (light grey) with foF2 = 5.4 MHz, compared to 4.3 MHz for the overhead ionosphere. The objective of the DPS development project was to develop alittle vertical incidence (i.e monostatic) ionospheric sounder which could automatically collect and analyze ionospheric measurements at remote operating sites for the aim of choosing optimum operating frequencies for obliquely propagated communication or radar propagation paths. Intermediate objectives assumed to be necessary to supply such a capability were the event of optimally efficient waveforms and of functionally dense signal generation, processing and ancillary circuitry. Since the necessity for an embedded general purpose computer was a given imperative, real-time control software was developed to include as many functions as was feasible into this computer instead of having to supply additional circuitry and components to perform these functions. The DPS duplicates all of the functions of its predecessor the Digisonde 256 [Bibl, 1981) and (Reinisch, 1987) during a much smaller, low power package. These include the simultaneous measurement of seven observable parameters of reflected (or in oblique incidence, refracted) signals received from the ionosphere:

1) Frequency
2) Range (or height for vertical incidence measurements)
3) Amplitude
4) Phase
5) Doppler effect and Spread
6) Angle of Arrival
7) Wave Polarization

Because the physical parameters of the ionospheric plasma affect the way radio waves reflect from or undergo the ionosphere, it's possible by measuring all of those observable parameters at variety of discrete heights and discrete frequencies to map and characterize the structure of the plasma within the ionosphere. Both the peak and frequency dimensions of this measurement require many individual measurements to approximate the underlying continuous functions.

An ionospheric sounder uses basic radar techniques to detect the electron density (equal to the ion density since the majority plasma is neutral) of ionospheric plasma as a function of height. The ionospheric plasma is made by energy from the sun transferred by particles within the solar radiation also as direct radiation (ultra-violet and x-rays). Each component of the solar emissions tends to be deposited at a specific altitude or range of altitudes and thus creates a horizontally stratified medium where each layer features a peak density and to a point, a definable width, or profile. The form of the ionized layer is usually mentioned as a Chapman function [Davies, 1989] which may be a roughly parabolic shape somewhat elongated on the highest side. The peaks of those layers usually form between 70 and 300 km altitude and are identified by the letters D, E, F1 and F2, so as of their altitude.

By scanning the transmitted frequency from 1 MHz to as high as 40 MHz and measuring the time delay of any echoes (i.e apparent or virtual height of the reflecting medium) a vertically transmitting sounder can provide a profile of electron density vs. height. this is often possible because the relative index of refraction of the ionospheric plasma depends on the density of the free electrons.

5.3.4 Magnetometer

A magnetometer may be a device that measures magnetism the direction, strength, or relative change of a magnetic flux at a specific location. The measurement of the magnetization of a magnetic material (like a ferrimagnet) is an example. A compass is one such device, one that measures the direction of an ambient magnetic flux, during this case, the Earth's magnetic flux.

The first magnetometer capable of measuring absolutely the magnetic field strength was invented by Carl Friedrich Gauss in 1833 and notable developments within the 19th century included the Hall Effect, which remains widely used.

Magnetometers are widely used for measuring the Earth's magnetic flux, and in geophysical surveys, to detect magnetic anomalies of varied types. In an aircraft's attitude and heading coordinate system, they're commonly used as a heading reference. Magnetometers also are utilized in the military to detect submarines. Consequently, some countries, like the us, Canada and Australia, classify the more sensitive magnetometers as military technology, and control their distribution.

Magnetometers are often used as metal detectors: they will detect only magnetic (ferrous) metals, but can detect such metals at a way larger depth than conventional metal detectors; they're capable of detecting large objects, like cars, at tens of metres, while a metal detector's range is never quite 2 metres. In recent years, magnetometers are miniaturized to the extent that they will be incorporated in integrated circuits at very low cost and are finding increasing use as miniaturized compasses

I. Magnetic fields

Magnetic fields are vector quantities characterized by both strength and direction. The strength of a magnetic flux is measured in units of tesla within the SI units, and in gauss within the cgs of units. 10,000 gauss are adequate to one tesla. Measurements of the Earth's magnetic flux are often quoted in units of nanotesla (nT), also called a gamma. The Earth's magnetic flux can vary from 20,000 to 80,000 nT counting on location, fluctuations within the Earth's magnetic flux are on the order of 100 nT, and magnetic flux variations thanks to magnetic anomalies are often within the picotesla (pT) range.[3] Gaussmeters and teslameters are magnetometers that measure in units of gauss or tesla, respectively. In some contexts, magnetometer is that the term used for an instrument that measures fields of not up to 1 millitesla (mT) and gaussmeter is employed for those measuring greater than 1 mT

II. Types of magnetometer

There are two basic sorts of magnetometer measurement. Vector magnetometers measure the vector components of a magnetic flux. Total field magnetometers or scalar magnetometers measure the magnitude of the vector magnetic flux. Magnetometers wont to study the Earth's magnetic flux may express the vector components of the sector in terms of declination (the angle between the horizontal component of the sector vector and magnetic north) and therefore the inclination (the angle between the sector vector and therefore the horizontal surface).

Absolute magnetometers measure absolutely the magnitude or vector magnetic flux, using an indoor calibration or known physical constants of the magnetic sensor. Relative magnetometers measure magnitude or vector magnetic flux relative to a hard and fast but uncalibrated baseline. Also called variometers, relative magnetometers are wont to measure variations in magnetic flux.

Magnetometers can also be classified by their situation or intended use. Stationary magnetometers are installed to a hard and fast position and measurements are taken while the magnetometer is stationary.Portable or mobile magnetometers are meant to be used while in motion and should be manually carried or transported during a moving vehicle. Laboratory magnetometers are wont to measure the magnetic flux of materials placed within them and are typically stationary. Survey magnetometers are wont to measure magnetic fields in geomagnetic surveys; they'll be fixed base stations, as within the INTERMAGNET network, or mobile magnetometers wont to scan a geographical area

Magnetometers specifically wont to measure the Earth's field are of two types: absolute and relative (classed by their methods of calibration). Absolute magnetometers are calibrated with regard to their own known internal constants. Relative magnetometers must be calibrated by regard to a known, accurately measured magnetic flux.

The simplest absolute magnetometer, devised, consists of a permanent magnet suspended horizontally by a gold fibre. Measuring the amount of oscillation of the magnet within the Earth's magnetic flux gives a measure of the field's strength.

A widely used modern absolute instrument is that the proton-precession magnetometer. It measures a voltage induced during a coil by the reorientation (precession) of magnetically polarized protons in ordinary water.

CHAPTER VI

RESPONSE OF IONOSPHERIC TOTAL ELECTRON CONTENT (TEC) OVER INDIAN ANTARCTIC STATION, MAITRI, 2008

6.1. Introduction

The ionosphere is the partially ionized region of the Earth's upper atmosphere. It extends from about 60 km to 1000 km. The main source of the ionization in the ionosphere is the solar radiations such as extreme ultra violate (EUV) and X- ray radiations. In addition to photoionization, collisional ionization due to particle precipitation from the magnetosphere is another source of ionization, in particular in the high latitude region. Once the plasma is produced by these processes, it undergoes chemical reactions with neutrals, diffuses due to the gravitational force and plasma pressure gradients, and is transported via neutral winds and electric fields under the influence of the Earth's magnetic field. Due to the altitude variations in the atmospheric neutral composition and the production rate with altitude, the plasma density in the ionosphere has a vertical layered structure, denoted by the D, E, and F layers (Figure 6.1). Each layer is controlled by different physical processes and has different main ions. In the D and E regions, the main ions are O_2^+, N_2^+, NO^+ and photochemistry is dominant. The F layer is usually divided into three sub-layers. The lowest layer, where photochemistry is dominant, is called the F1 layer. Here the ionization is produced through the photoionization process and disappears through recombination processes with the electrons. The next sub-layer where the transition from chemical to diffusion occurs is called the F2 layer, there are maximum electron density usually occurs. The uppermost part of the ionosphere, above the F2 layer is termed the topside of the ionosphere. Here diffusion dominates (Schunk and Nagy, 2000). In addition to the variation of the plasma density with altitude, the ionosphere also shows significant variations with time of day, latitude, longitude, season, solar activity, and geomagnetic activity. A distinctive latitudinal characteristic in the ionosphere is created owing to the geometry of the Earth's dipolar magnetic field lines.

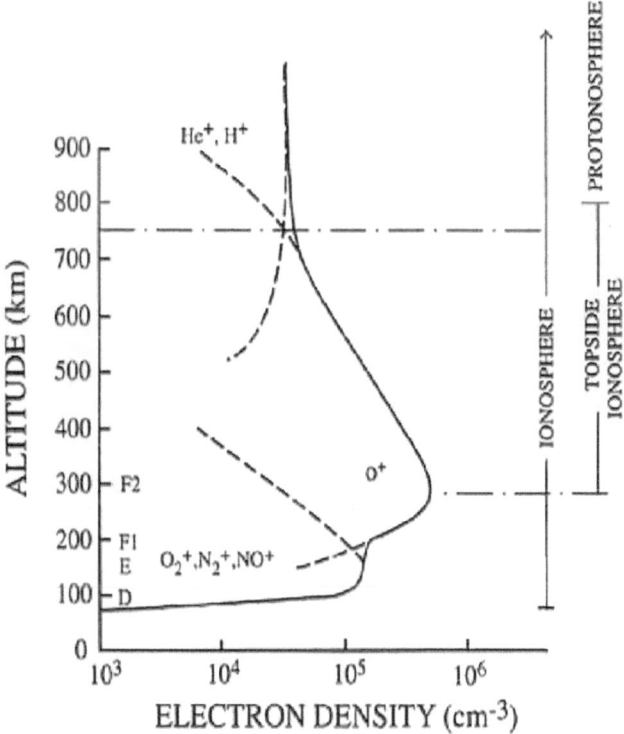

Figure 6.1: Ion density profiles for the daytime middle-latitude ionosphere showing the layered structure (D; E; F1, and F2) [Banks et al., 1976].

The ionosphere, therefore, is classified into three latitude regions, low (equatorial), middle, and high (auroral) latitude regions, which are controlled by different physical processes. Radiation from the sun causes ionization in the ionosphere. Electrons are produced when this radiation collides with uncharged atoms and molecules (Figure 6.2). Since this process requires solar radiation, production of electrons only occurs in the daylight hemisphere of the ionosphere.

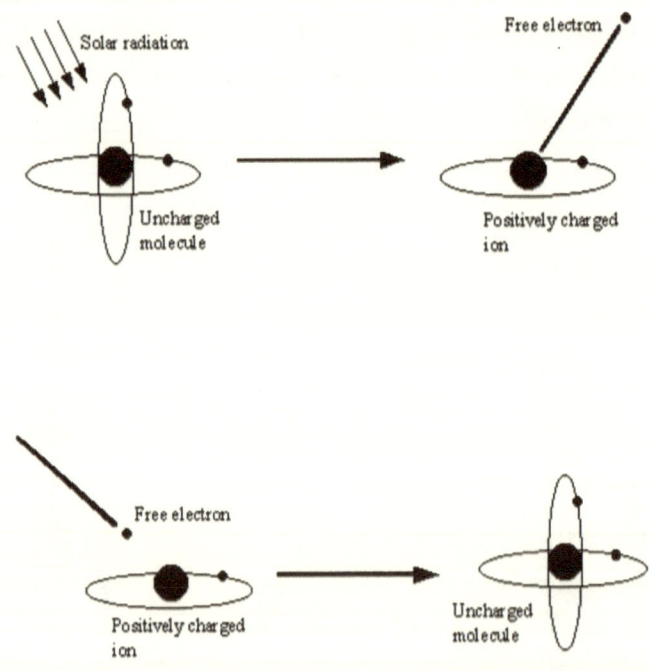

Figure 6.2: Production (top) and loss (bottom) process in the ionosphere (Image source by Nova Science Centre)

When a free electron combines with a charged ion a neutral particle is usually formed (Figure 6.3). Essentially, loss is the opposite process to production. Loss of electrons occurs continually, both hemispheres.There are three major regions of the global ionosphere. These are the high-latitude, midlatitude and equatorial or low latitude regions. In this section, we briefly described the main characteristics of the individual regions.

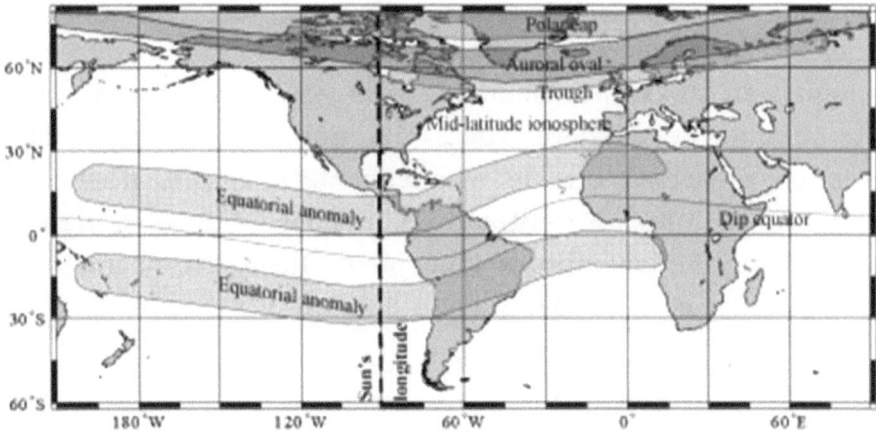

Figure 6.3: Geographic classification of the ionosphere
(source by Space weather websites)

I. The Low-Latitude or Equatorial Ionosphere

At low latitudes, during the day, one of the most prominent features in the ionosphere is the equatorial anomaly, which is also often called the Appleton anomaly (Appleton, 1946). This feature is eminent by higher plasma density on both sides of the equator, rather than at the equator itself. The equatorial anomaly is formed as a consequence of ExB upward plasma drifts connected with an eastward E electric field and a northward horizontal B magnetic field. The lifted plasma then diffuses downward along the geomagnetic field lines due to the gravitational force and the plasma pressure gradient, and this results in the ionization enhancements on both sides of the magnetic equator (at about $\pm 10^0 \sim \pm 20^0$ in the latitude). This physical mechanism phenomenon is called the fountain effect (Figure 6.4). Often, irregularity is found between the northern and southern anomaly. Due to an interhemispheric wind blowing from the summer to the winter hemisphere, in the summer hemisphere, plasma moves upward along the geomagnetic field lines, while plasma moves downward in the winter hemisphere. Therefore, the transport of the lifted plasma toward the winter hemisphere is enhanced, and the plasma transport toward the summer hemisphere is decreased. As a result, the equatorial anomaly in the winter hemisphere is generally larger than in the summer hemisphere. The Equatorial anomaly morphology is often disturbed due to magnetic storm effects (Rishbeth, 1975; Abdu, 1997). Recently, there has been a growing interest in studying the low-latitude ionosphere during major magnetic storm (Kelley et al., 2004; Lin et al., 2005).

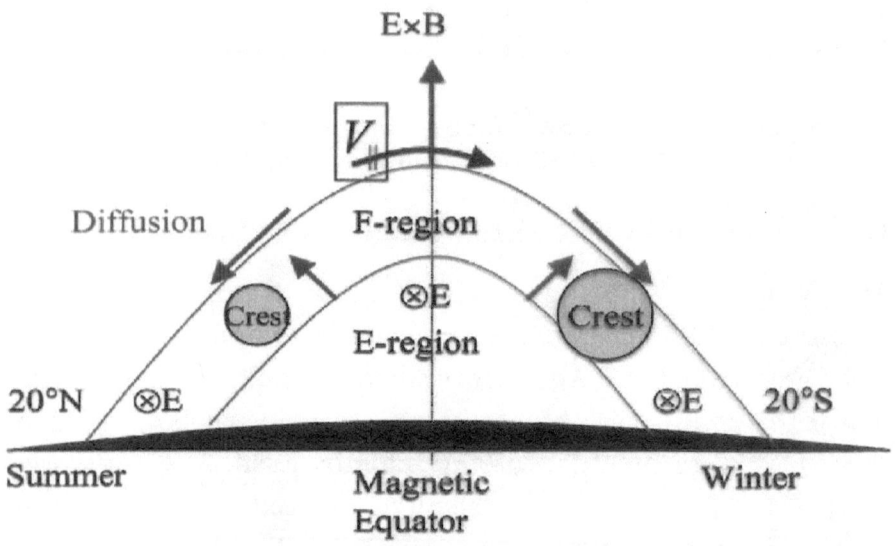

Figure 6.4: Fountain effects over the equatorial regions. An interhemispheric wind blowing from the summer to the winter hemisphere produces an asymmetry between two peak densities of the equatorial anomaly. E denotes an eastward electric field and B is the northward geomagnetic field.

II. The Middle-Latitude Ionosphere

The middle-latitude ionosphere is known to be the best understood ionospheric region, largely due to the relatively simple physics and the reasonably good coverage of measurements. In the middle-latitude ionosphere, transport of plasma is mostly caused by the combination of two facts, (i) the geomagnetic field line is inclined to the horizontal ii) the ionospheric plasma is constrained to move along the geomagnetic field lines. Therefore, thermosphere neutral winds effectively transport the plasma along the field lines into higher or lower altitude regions in which recombination rates are different resulting in changes of the plasma density. During the day, the typically pole ward neutral winds move plasma down to lower altitudes where the recombination rate is large. This results in a reduced peak height of F2 and a decrease in the peak electron density. On the contrary, during the night time, the typically equatorial winds, move plasma up. Therefore the recombination of the plasma with neutrals decreases, the peak height increases, and the nighttime peak electron density is maintained.

III. The High -Latitude Ionosphere

In addition to photon ionization, coalitional ionization is another source of ionization in the high-latitude region. The main reason for this is the fact that the geomagnetic field lines are nearly vertical in this region leading to the charged particles descending to E layer altitudes (about 100 km). These particles can collide with the neutral atmospheric gases causing local enhancements in the electron concentration, a phenomenon which is associated with auroral activity. Auroral activity can also be regarded as an interaction between magnetosphere, ionosphere, and atmosphere. The auroral zones are relatively narrow rings situated between the northern and southern geomagnetic latitudes of about 64 and 70 degrees. In general, the intensity and the position of the auroral ovals are related to geomagnetic disturbances. The ovals expended towards the equator with increasing level of geomagnetic disturbances (McNamara, 1991).

On the equatorial side of the auroral ovals lines the mid-latitude through which is a narrow region of the ionosphere with a width of a few degrees. It can be characterized by a sudden drop in the critical frequencies and electron densities by a factor of two or more. This occurs essentially at night time primarily due to the increased recombination as a consequence of the shorter high latitude day time ionization periods (Tascione, 1988).

The direct interaction between the magnetosphere and the interplanetary magnetic field results in the dayside cusp or cleft. It is typically 2 to 4 degrees wide and located at the geomagnetic latitude of 78 to 80 degrees near local noon. The phenomenon can be characterized with enhancements in electron densities at all altitudes.

The geographical regions enclosed by the auroral rings are called the polar caps. Our understanding of the polar cap region is rather limited due to the lack of available information. The polar caps are largely affected by solar flares and corona mass ejections from coronal holes (relatively cool "open" structures of the solar corona) causing D region electron density enhancements.

6.2. Methodology for Measurements

One of the widely used ionospheric parameter is Total Electron Content (TEC), which is the number of electrons in a column of one-meter-squared cross section that extends all the way up from the ground through the ionosphere. TEC can provide not only an overall description of the ionization in the ionosphere, but also can be used for practical applications of radio wave propagation. Single frequency Global Positioning System (GPS) users can use TEC measurements to correct their signal, since TEC is proportional to the radio signal delay that a GPS signal experiences in the ionosphere. In this section, the effects of the ionosphere on radio wave propagation and the TEC measurement methods will be described.

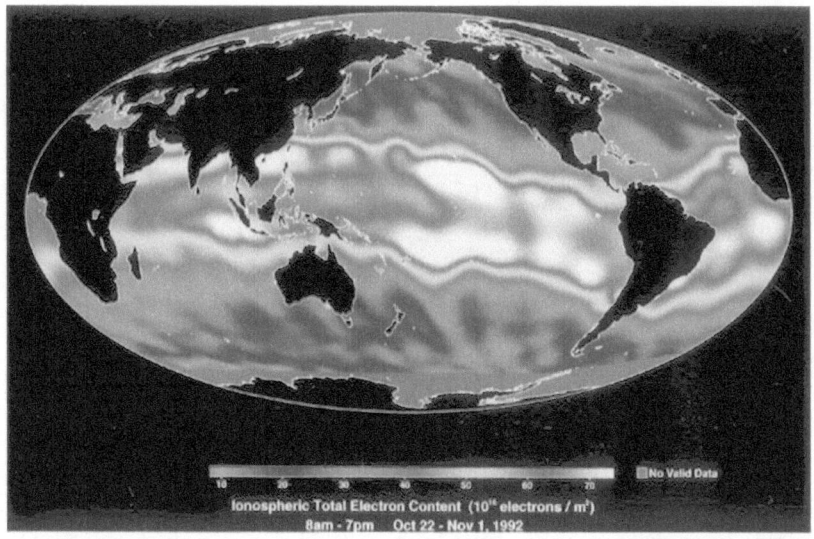

Figure 6.5: Global Ionospheric Total Electron Content Map

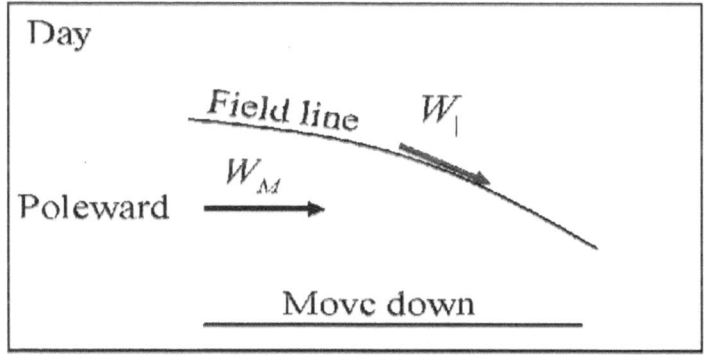

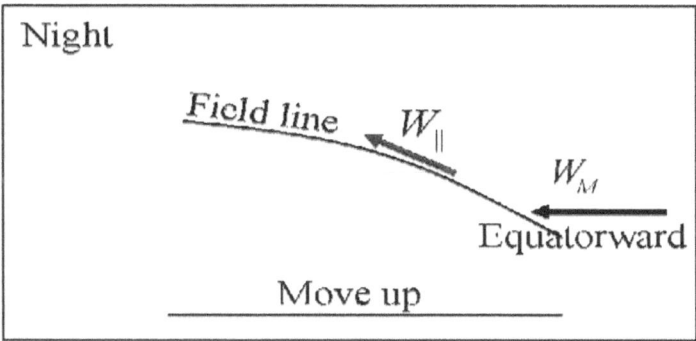

Figure 6.6: Vertical plasma drifts due to the meridional neutral wind W_M. $W\|$ is the meridional wind along the geomagnetic field line.

Behavior of polar ionosphere is entirely different from the equatorial or low latitude ionospheric behavior since it is dominated by high geomagnetic activity and receives an entirely different radiation budget. The structure of the high latitude ionosphere is very complicated, varied up on and exhibits a very typical behavior because the Earth's magnetic field focuses many geospace effects to these regions. One of the most interesting characteristic features of the polar regions is the auroral activity. The energy transferred from solar wind to magnetosphere during solar wind magnetosphere interaction is being deposited at the polar region through reconnection process which makes polar ionosphere highly irregular. Higher latitude regions are directly affected by the entry of charged solar particles which on dissipating their energy cause auroras and thus make the ionosphere highly irregular.

The auroral ionosphere is subject to adverse space weather conditions which cause significant temporal and spatial variations of electron density and the density gradients resulting in highly variable Total Electron Content (TEC) (Aarons et al., 2000) and fluctuation of radio signal amplitude and phase (Basu et al., 1998). In high latitude region, irregularities at a different scale are common, which causes fluctuations in the total electron content. The intensive phase fluctuations observed along GPS satellite passes are caused by dramatic changes in Total Electron Content (TEC) and demonstrate a strong horizontal gradient of TEC. Fluctuation effects and TEC gradients can have a different impact on GPS measurements. They affect phase ambiguity resolution, increase the number of undetected and uncorrected cycle slips and loss of signal lock (Wanninger, 1993; Krankowski et al., 2002).

Several studies (Coker et al., 1995; Aarons, 1997: Aarons et al., 2000; Krankowski et al., 2005) have used GPS observations from a single site or local network to monitor TEC fluctuations and related irregularities in the high latitude ionosphere. Wanninger (1995) and Doherty et al., (1994) have utilized the GPS data at 30 s interval to study ionospheric irregularities of electron density by computing the time rate of the change of the differential carrier phase. This is equivalent to the rate of the change of the total electron content, termed ROT in units of TEC/min. Aarons (1997) and Krankowski et al. (2005) have demonstrated the utility of such a dataset for studying the evolution of different scale irregularities during magnetic storms at high latitudes. Pi et al. (1997) have defined a rate of change of TEC index (ROTI) based on standard deviation of ROT over a 5-min period. Annual and semiannual VTEC effects at low solar activity based on GPS observations at different geomagnetic latitude also studied by Natali et al. (2010).

Over past four decades, a great deal of research has revealed that ionospheric scintillation is most likely to occur in equatorial and auroral regions. At low latitudes, the scintillation is primarily controlled by increasing irregularities over the magnetic equator. After sunset, when the eastward electric field is enhanced, irregular plasma density depletions are generated on the bottom side of the nighttime equatorial F region and rises to higher altitudes as a result of nonlinear evolution of the generalized Rayleigh-Taylor and ExB instabilities (Basu et al., 1978; Kelley 1989; Fejer et al., 1999). In the auroral zone, scintillations mainly occur in the nighttime period and exist at all local time in the polar cap region. In recent years many observations of GPS scintillations at high latitudes were reported by many researchers (De Franceschi et al., 2006; Meggs et al., 2008). Using GPS observations from 11 high latitude stations, Aarons (1997) noted that phase fluctuation activity has a daily pattern mainly controlled by the motion of the receiver location into the auroral oval. Mitchell et al., (2005) found that GPS amplitude and phase scintillation collocated with steep Total Electron

Content (TEC) gradient at the southwest of Svalbard during the Hallowean storm of October 2003. Later, De Franceschi et al., (2008) examined the observations from a chain of GPS resources in Northern Europe, and investigated the dynamics of ionospheric scintillation and TEC during the storm event. A strong influence of IMF on the formation and movement of patches was reported.

6.3. Results and Discussion

For the present study we have used the data which was collected by Indian Antarctic Scientific Expedition at Indian base stations from January 2008 to January 2009 (Maitri, Antarctic). Ionospheric scintillation and TEC measurement were performed using the GPS Ionospheric Scintillation and TEC monitors (GISTM) GSV 4004A. GISTM consists of a NovAtel OEM4 dual frequency receiver with special firmware specifically configured to measure log power and phase of the GPS L1 signal at high sampling rate (50 Hz). The receiver computes ionospheric TEC from the GPS L1 and L2 signals. The GSV 4004A can also automatically compute and record the amplitude scintillation index, S4, which is the standard deviation of the received power normalized by its mean value, and the phase scintillation index S4, the standard deviation of the detrended phase using a filter in the receiver with 0.1 Hz cutoff. This receiver is capable of tracking and reporting scintillation and TEC measurements simultaneously up to 10 GPS satellites in view (Jayachandran et al., 2009). The GPS observables are biased by the instrumental delay therefore it is necessary to remove these biases for accurate estimation of TEC. The absolute Total Electron Content (VTEC) determination has the capability to remove the instrumental biases both from the receiver and the satellite. The instrument time delay and potential errors are corrected using the code biases.

A total of one year of GPS-TEC data have been processed for Indian permanent station Maitri, Antarctic during the year of 2008. The study divided in three parts:-

6.4. Monthly behavior of Total Electron Content:-

The observation for monthly variation of TEC is based on 12 months GPS data. During the month of January TEC fluctuated between the ranges of 10 to 22 TECU. In the February month 7 to 22 TECU, March month TEC fluctuated between 6 – 20 TECU. In the April month minimum TEC observed as 6 TECU and maximum 15 TECU observed. In the starting of polar night moth May, minimum TEC as 7 TECU and maximum 15 TECU observed, in a Dark month of June minimum TEC 4 TECU and maximum 15 TECU observed, in the month July minimum TEC goes to 3 TECU and maximum TEC 16 TECU observed, in the starting of suns activity month August minimum TEC observed 3 TECU and maximum 19 TECU, in the spring month of Antarctica September and October minimum TEC noted 3 and 7 TECU and maximum 22 and 22 TECU, again in the summer month November and December TEC variation observed between the range of minimum 6 and 8 TECU and maximum 27 and 26 TECU. This type of behavior of TEC in polar region depends on solar zenith angles. Figure 6.7 clearly shows 12 month TEC behavior in all months and we noted in every month TEC pick shift pattern between the month January to May TEC peak sifted right side, between the month of June and July TEC peak almost observed overlapped but again sun rising month August to peak summer month December TEC peak noted toward left side. This type of peak shifting pattern depends on the solar zenith angle.

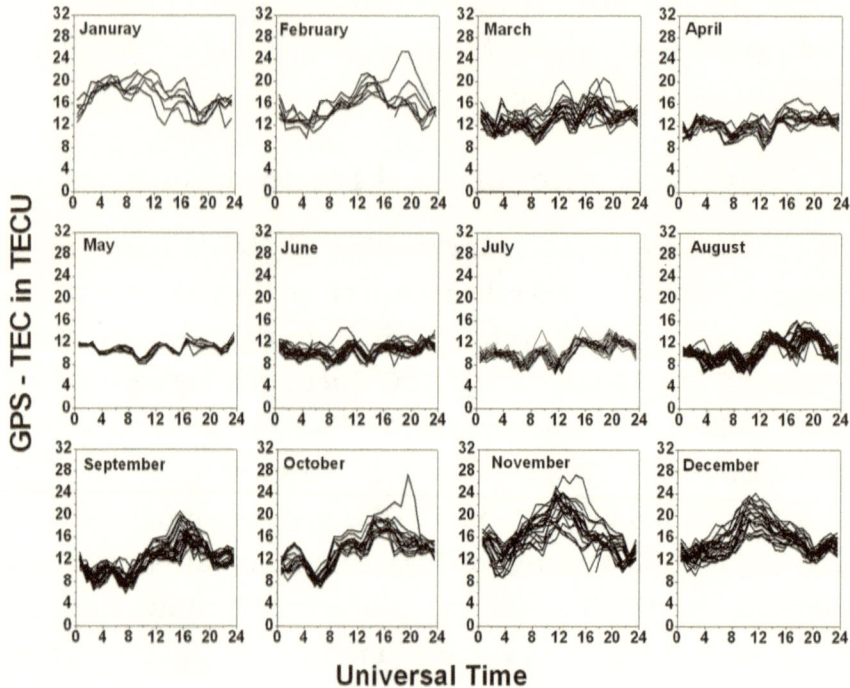

Figure 6.7: Monthly Total Electron Content variations over Maitri, Antarctic

6.5. Seasonal Variation of Total Electron Content (TEC)

The observation for seasonal variation of ionospheric TEC over Antarctic is divided into three seasons: first is summer season (November, December, January and February) Second is winter season (May, June, July and August) and third is equinox period divided into two equinoxes seasons according to solar activities first Autumnal equinox (March, April) and other Vernal equinox seasons (September and October). In the summer period TEC monthly median value fluctuate in the range of 11 – 20 TECU, this type of performance of TEC in summer period cause of presence of 24-hour solar activity in polar region. In the period of winter TEC monthly median value drops and fluctuates between the ranges of 8 to 14 TECU. During the winter period solar activities are negligible as compared to summer period. The study of equinox period is divided into two parts first is Autumnal equinox from March to April and second is Vernal equinox from September to October 2008. In the autumn equinox, the monthly median value of TEC fluctuate between the ranges of 8-16 TECU, it is because of partial decrease of solar activities. In other vernal equinox period TEC fluctuates between 8-18 TECU during the period of partial increase in solar activities. Figure 6.8, shows seasonal variation of TEC over Maitri, Antarctic.

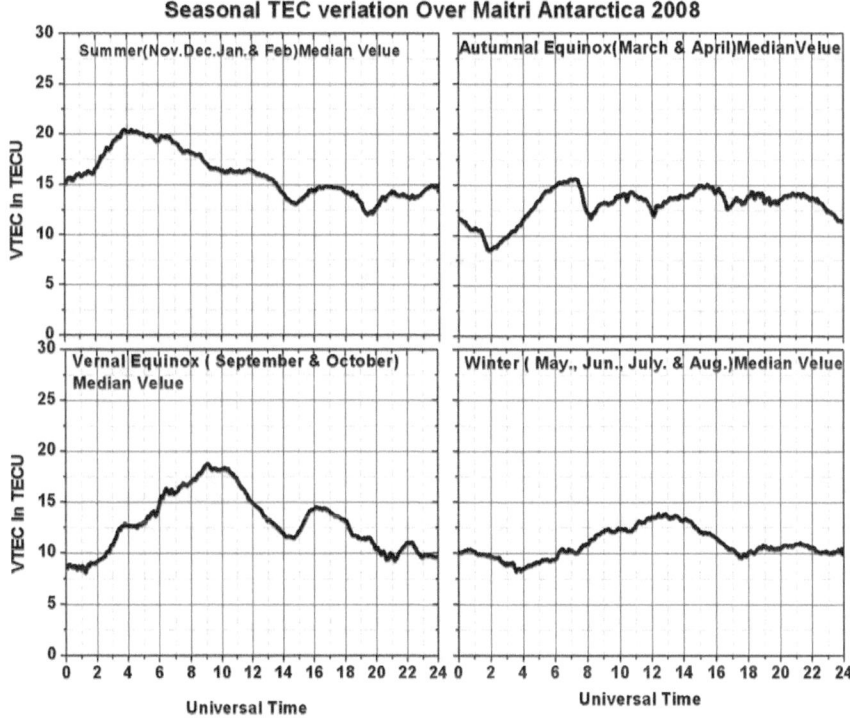

Figure 6.8: Seasonal Variations of Total Electron Content at Maitri, Antarctic

6.6. Polar Day and Polar Night Observation

The observation during the mid polar night (21 June 2008) is as shown in Figure 6.9. This month is totally Antarctic dark night month, during this night, total solar activities are absence and TEC variation is very less, TEC behavior is just apposite during mid polar day (21 December 2008). In this month complete full sunny days and therefore high solar activities are present, and TEC fluctuations are highest as compared to polar night. Figure 2.7 shows mid polar day and mid polar night variations.

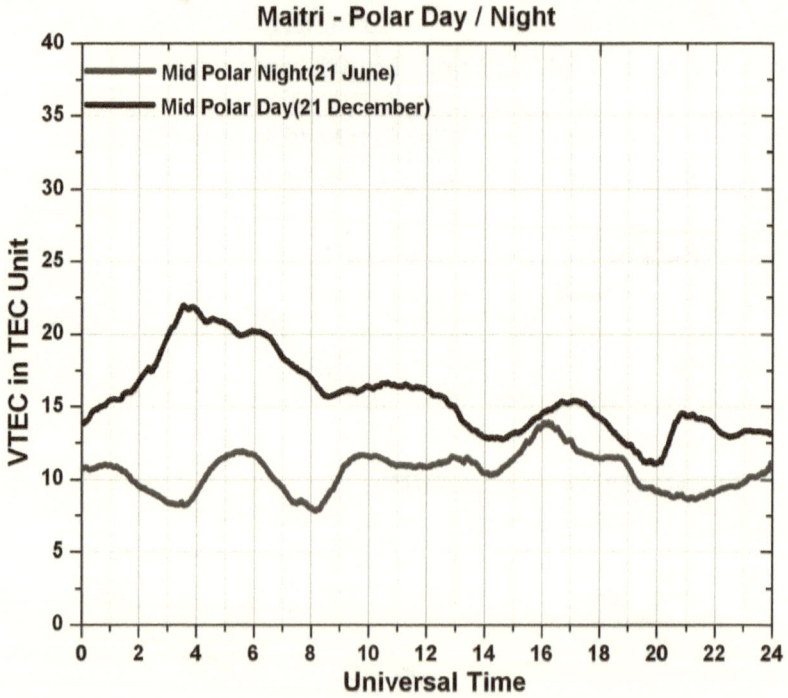

Figure 6.9: Mid polar day and mid polar night variation of Total electron Content at the Maitri, Antarctic

6.7. Conclusions

We have processed and analyzed one year GPS data from Indian base station Maitri, Antarctic, present work shows the presence of strong relationship between GPS-TEC and its dependence on solar activities. During the Monthly observation in the first January month TEC value goes to high as compared to June month, TEC value again goes to high in December month. In

January and December months solar activities are high as compared to June month. The seasonal study of Total Electron Content (TEC) shows that in the winter season total solar activities are discontinued and no ionization processes during this season so that TEC value goes to minimum. The study of summer season just is a opposite of winter season, the total solar activities are present which is due to ionization process in presence of 24 hour sun light than TEC value goes to maximum. In our last study of equinox season which is divided into two parts: in the first equinox period (March and April months) solar activities are low due to maximum solar zenith angle, which is the time of the sunset. In the second equinox period (September and October months) solar zenith angle goes toward maximum to minimum and solar activities are again started and ionization process progresses and during this period TEC variation are high as compared to March and April months. Ionospheric Total Electron Content (TEC) variation depends on solar activities and solar zenith angle and TEC variation are high in presence of solar activities and minimum solar zenith angles and TEC variation are low in absence of solar activities and maximum solar zenith angle.

CHAPTER VII

IONOSPHERIC SCINTILLATION BEHAVIOR AT INDIAN ANTARCTIC STATION MAITRI, 2008

7.1 Introduction

The high latitude ionosphere often remains turbulent and develops electron density irregularities due to solar flares-magnetosphere-ionospheric interactions. The dimensions of these high latitudinal ionospheric irregularities range from few meters to kilometers and cause the GPS signals to scatter in terms of amplitude and phase commonly referred to as scintillations. These types of scintillations are also known as auroral scintillations. The ionosphere is a dispersive medium in which RF signals are refracted by the amount dependent on the signal frequency and electron density in regions of irregularities resulting in random phase variations in the emerging wave front known as phase scintillations. Diffraction of signals also leads to variations in signal amplitude referred to as amplitude scintillation (or amplitude fading for degradations in signal strength). These effects are strongest in equator, auroral and polar cap regions. High latitude auroral irregularities are formed from the precipitation of energetic electrons along terrestrial magnetic field lines into the high latitude ionosphere. These electrons are energized through complex interaction between the solar wind and the earth's magnetic fields, resulting in optical and UV emissions commonly known as the auroras. This phenomenon characterizes the magnetospheric substorm, where associated irregularities in electron density lead to scintillations (Arons, 1982). At high latitudes scintillations are found to be associated with large scale plasma structures. Experimentally, the two states of the polar ionosphere controlled by the IMF, and their association with high latitude large scale plasma structures known as patches, blobs and sun–aligned arcs, have been discovered in the 1980's (Weber et al., 1984,1986, Tsunda,1988, Basu and Valladares,1999). It is well known that ionospheric scintillation is produced by electron density irregularities in the ionosphere, which becomes highly disturbed at times. A radio wave crossing these drifting ionospheric irregularities suffers a distortion of phase and amplitude and the magnitude of fluctuations vary with the frequency used, magnetic and solar activity conditions, and time of the day, season and location. Severe amplitude fading and strong phase scintillation affect the reliability of GPS navigational systems and satellite communications. Therefore, it is desirable to obtain further understanding of ionospheric scintillation and its effects on GPS by means of a receiver capable of performing in such conditions.

Ionospheric scintillation, which is produced by ionospheric irregularities, affects GPS signals in two ways, broadly classified as *refraction* and *diffraction*. Both types of effects originate in the group delay and phase advance that a GPS signal experiences as it interacts with free electrons along its transmission path. This chapter describes amplitude scintillation measurements over Indian Antarctica station Matri (Lat. 70°.65 S, Long.11°.45 E) as part of International Polar Year (IPY) by using Novatel dual frequency GPS receiver.

I. Ionospheric Scintillation

Ionospheric scintillation is a rapid fluctuation of radio-frequency signal phase and amplitude, generated as a signal passes through the ionosphere. Scintillation occurs when a radio frequency signal in the form of a plane wave traverses a region of small scale irregularities in electron density. The irregularities cause small-scale fluctuations in refractive index and subsequent differential diffraction (scattering) of the plane wave producing phase variations along the phase front of the signal. As the signal propagation continues after passing through the region of irregularities, phase and amplitude scintillation develops through interference of multiple scattered signals. Figure 7.1

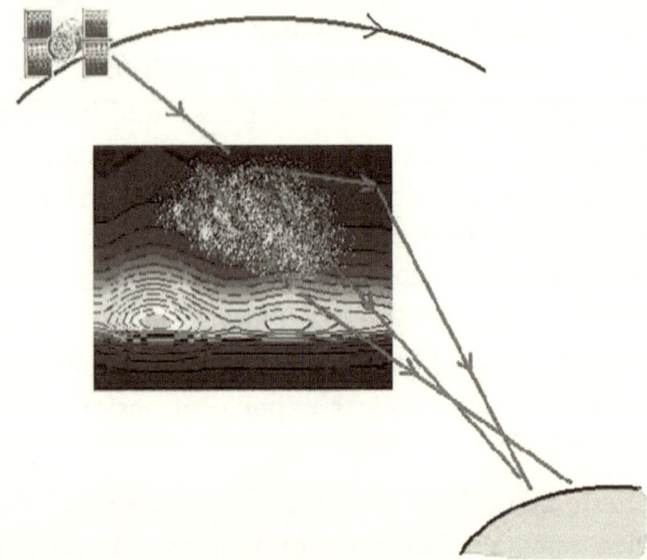

Figure 7.1: *ionospheric scintillation (source from NWRA Space Weather website)*

Ionospheric scintillation is a well-known phenomenon that has been studied extensively in the past yet it remains a difficult phenomenon to predict or model on a large scale. Scintillation is caused by small-scale fluctuations in the refractive index of the ionospheric medium which in turn are the result of inhomogeneities. Inhomogeneities in the ionospheric medium are produced by a wide range of phenomena (eg. plasma bubbles), and those responsible for scintillation occur predominantly in the F-layer of the ionosphere at altitudes between 200 and

1000km. The primary disturbance region, however, is typically in the F-region between 250 and 400km. E-layer irregularities such as sporadic-E and auroral E can also produce scintillation but their effect on L-band GPS signals is minimal.

Ionospheric scintillation is primarily an equatorial and high-latitude ionospheric phenomenon, although it can (and does) occur at lower intensity at all latitudes.

In terms of geographic (geomagnetic) distribution, ionospheric scintillation generally peaks in the sub-equatorial anomaly regions, located on average ~15° either side of the geomagnetic equator, as can be seen in Figure 7.2.

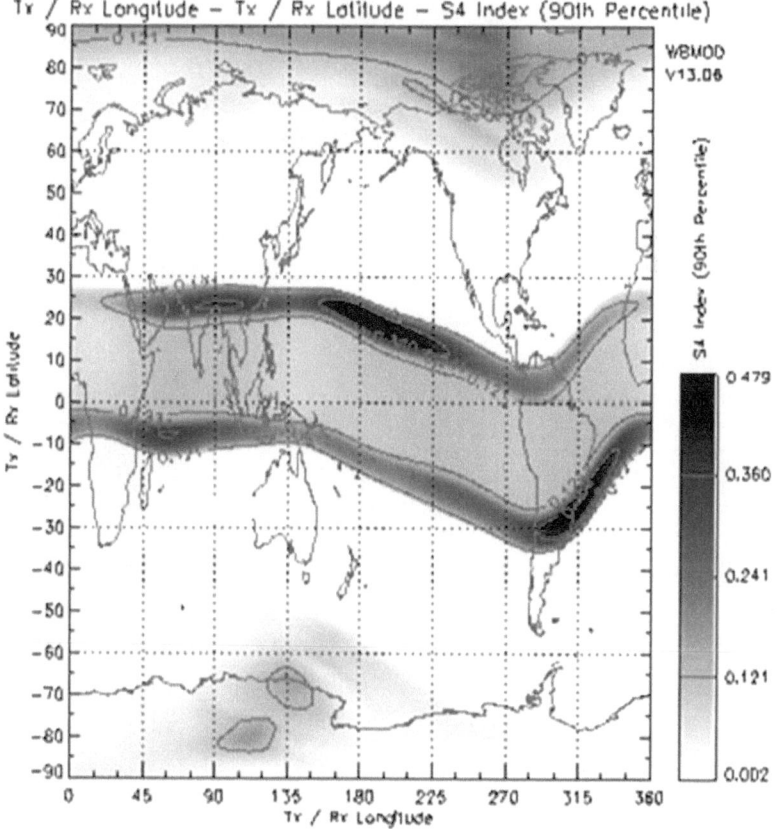

Figure 7.2: S4 Scintillation index at GPS L1 (1575.42MHz) assuming constant local time (=2300) at all longitudes. (Source by http://www.ips.gov.au/Satellite/6/3)

The figure shows "WBMOD" model predictions of the 90th percentile S_4 index at 2300 Local Time (everywhere) at the S.Hem autumnal equinox (DOY 091) for GPS L1 (1575.42MHz), low magnetic activity (Kp=10) and high solar activity (SSN=150). Apart from the two strong scintillation bands following ~15° geomagnetic latitude contours, also obvious is the enhanced scintillation between the two bands of maxima and in the Polar Regions. The mid-latitude regions are relatively free of scintillation, especially at GHz frequencies, however at lower frequencies, closer to 100MHz there can at times be significant scintillation activity. In terms of diurnal distribution, equatorial ionospheric scintillation generally peaks several hours after dusk.

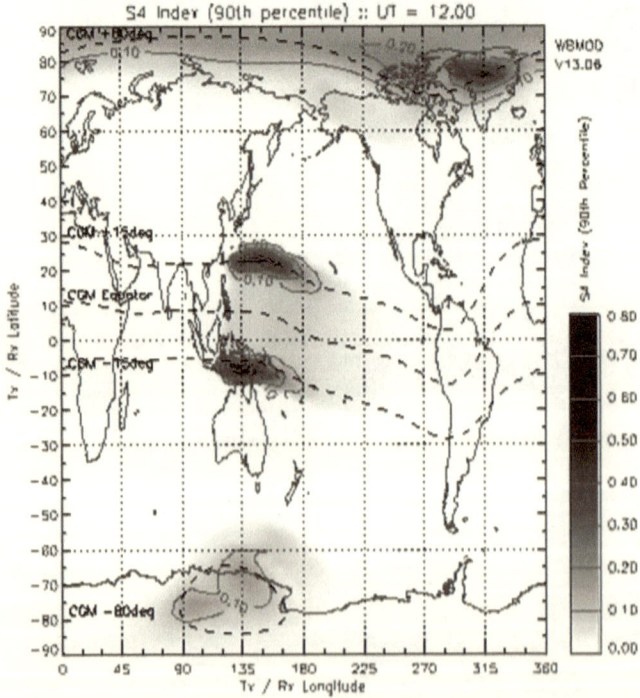

Figure 7.3: S_4 Scintillation index at GPS L1 (1575.42MHz) assuming constant Universal time (=1200). The dashed lines represent lines of constant geomagnetic latitude (Source by http://www.ips.gov.au/Satellite/6/3).

The figure shows "WBMOD" model predictions under the same conditions as Figure 7.3 but for 1200UT, rather than at constant local time. The choice of 1200UT during equinox means the left and right hand borders of the plot are at midday local time, the vertical centre line of the plot (longitude 180°) corresponds to local midnight, and dusk is at longitude 90°. Each division on the X axis (15°) corresponds to 1 hour. Again, two strong scintillation bands can be seen roughly corresponding to ±15° geomagnetic latitude (as indicated by the dashed lines). The equatorial scintillation is present at decreasing intensity levels throughout most of the nightside. The scintillation peak in the equatorial regions occurs between 2100 and 2200 local time.

In general, GPS receivers located at mid-latitude sites in Australia (-40° < latitude < -20°) will not be significantly affected by ionospheric scintillation. GPS receivers located at latitudes north of -20° in Australia may experience some degree of ionospheric scintillation, primarily when tracking satellites at low elevation angles to the north, and during active geomagnetic conditions. Locations as far north as Darwin and the Cape York Peninsula are likely to experience ionospheric scintillation more regularly on satellites tracking to the north of the station since the Line Of Sight (LOS) to these satellites will generally pass through the equatorial anomaly regions. GPS receivers located at latitudes southwards of -40° will commonly see ionospheric scintillation activity associated with geomagnetic storm activity, on satellites tracking to the south of the receiver, with the most significant scintillation occurring under the auroral zones.

II. Characteristics of Scintillations

The properties and processes of mid-latitude ionospheric scintillations, collectively known as the morphology, is vital to any intensive effort aimed at examining the likely occurrence of scintillations of transionospheric signals. Drawing on over fifty years of research effort in the area, it can be seen that the mid-latitude region has unique characteristics that must be taken into account when engaged in experimental observations and postulating theories. While it is true that the general theories of ionospheric scintillation are valid for mid-latitudes, a recognition exists that the processes leading to the formation of electron density irregularities in geomagnetic mid-latitudes are often different from those of the high and low latitude regions. This recognition then places an onus on investigators to be aware of the position of an irregularity with respect to its latitudinal classification, thereby removing the possibility that the scintillation activity is attributed to a process belonging to another latitudinal range.

In order to understand the situation fully, it is first necessary to present a clear definition of the extent of the mid-latitude region. This presents a problem in that the mid-latitude region is defined, by exclusion, to be the region of the Earth's ionosphere between the high and low latitude ranges. The boundaries of the high and low latitude regions of the ionosphere can be uniquely distinguished by examining several indicators but it is important to note that these boundaries are not fixed in time, showing diurnal, seasonal and other periodicities. It is sufficient to outline the geomagnetic latitudinal ranges that shall pertain to the labels of high, low and mid-latitudes.

+ high-latitudes: ± 60°- 90°
+ mid-latitudes: ± 20°- 60°
+ low-latitudes: ± 0°- 20°

These ranges are not arbitrarily defined but are selected as the regions in which specific ionospheric processes take place. The low latitude region is the sector in which equatorial ionospheric effects are the most important, such as the equatorial electrojet, whilst the high latitude region covers those areas in which polar and auroral ionospheric effects are noteworthy eg. high energy particle precipitation. The absences of such effects, which induce severe perturbations to the quiet ionosphere, serve as the defining characteristic of the mid-latitude ionosphere. The mid-latitude ionosphere is the quietest region and it can be noted that such events as scintillation of trans-ionsopheric signals are greatly reduced in magnitude from their counterparts in the other regions. Although less complicated than the high and low latitude cases, the morphology of mid-latitude ionospheric scintillations is not as simple as is sometimes stated, especially when effects from the high and low latitudes spill over into the mid-latitude ionosphere.

III. Types of Ionospheric Scintillation

Hajkowicz (1994), in a review of scintillations over a sunspot maximum, presents a classification system that distinguishes between ionospheric irregularities originating in the equatorial, auroral and mid-latitude regions viewed by observation posts located in the southern mid-latitude range. Adopting this system as a convenient labelling criterion, we are presented with three types of scintillation:

+ N type
+ P type (subset Ps type)
+ S type

As pointed out in Hajkowicz's paper, for each of the scintillation types, the *"occurrence pattern varies depending on the time of day, season and solar cycle."* Observing this structure, the morphology of the scintillations is discussed most accurately by analyzing each type in detail. It would seem prudent to do so by discussing their occurrence, relations to various periodic cycles, production by ionospheric disturbances of specific kinds and in terms of a number of analytical measures, such as spectral index. Before this process can take place, a clear definition for each type should be presented, preferably with reference to some latitudinal distinction. Consideration must also be given to scintillations that display unique features in the records, such as quasi-periodic or QP scintillations, which while indistinguishable in morphology from random scintillations, require additional care in the analysis of the propagation geometry and other factors.

N type scintillations: - N type scintillations are defined to occur in the vicinity of the pole wards boundary of the equatorial scintillation region. (In the case of Hajkowicz's study, this corresponded with the sub-ionospheric points to the north of his stations and hence his use of the designation N). N type scintillations have a maximum in their occurrence during periods of heightened solar activity when the equatorial scintillation belt creep pole wards as the equatorial electrojet intensifies. Heightened activity is found to occur at sunspot maximum and a seasonal periodicity is observed, peaking for the summer-autumn period (in Hajkowicz's study this was for the austral summer-autumn period.) N type scintillations are more frequently observed at night although increased activity is often observed during the summer daytime. This daytime maximum is closely correlated with the well known increase in E_s occurring in the summer daytime.

The observation of N type scintillations is strongly affected by the position of the receiver location and the geometry of the antenna. In true mid-latitudes, for observations confined to small zenith angles (less than 45°), N type scintillations are rarely observed. If however one were to consider the case of small elevation angles, N type scintillation is likely to be observed frequently. The observation of geostationary satellites from mid-latitude regions must therefore take into account the position of the equatorial scintillation oval when examining the causative mechanism behind the scintillation occurrence. For specific geometrical alignments, such studies are susceptible to the enhancement of scintillation phenomena associated with the aspect angle sensitivity discussed by Sinno and Minakoshi (1983).

P type scintillations: - P type scintillations are so called because the electron density irregularities responsible for the disturbance to the transionospheric signal occur in confined patches within the mid-latitude ionosphere. The patches have been found to consist of rod-like field-aligned irregularities (FAIs) associated with disruption to the F2 layer in the form of range spread-F. As the only true mid-latitude ionospheric scintillations, it is important to note that their specific morphology is quite different to that of the N and S types. While the N and S types occur more frequently in the perturbed ionosphere during sunspot maximum, the P type scintillations are noted to occur within the quiescent night-time ionosphere of geomagnetic mid-latitudes during sunspot minimum. This pattern of behaviour is strongly correlated with the maximum in spread-F occurrence during the same time frame, as commented on by (Bowman and Hajkowicz, 1990) and others. A subsiduary maximum can also be found during sunspot maximum during daylight hours which is thought to be associated with sporadic-E, although this enhancement is relatively minor. A subset type PS, where the patches lie to the south of the station away from magnetic zenith, demonstrates similar occurrence patterns as the main P type.

S type scintillations: - S type scintillations are found to be associated with the formation of irregularities in the vicinity of the equatorwards edge of the auroral scintillation belt. S type scintillations commonly produce the most intense scintillations recorded at mid-latitude stations. The occurrence maximum for this type of scintillation is found to show a close correlation with the maximum in the 10.7 cm solar radio flux that takes place for solar maximum. This enhancement peaks during the sunspot maximum which reflects the migration of the boundary of the auroral scintillation oval to lower latitudes during heightened solar activity. A seasonal dependence with maximum during the austral summer-autumn period is noted to take place, coupled with a diurnal enhancement in the early evening and mornings when the ionosphere is perturbed by effects associated with the sunrise and sunset terminator. The incidence of S type scintillations during the summer daytime is found to coincide with the well known pattern of occurrence for sporadic-E. The minimum occurrence of this type of scintillation is observed during solar minimum during the winter. Most scintillations of this type occur during the night-time. No significant increase in scintillation activity can be found for periods with larger values of the average planetary magnetic index A_p, although sensitive dependence is observed during the equinoctial months of September and March.

QP scintillations: Quasi-periodic (QP) scintillations are so named because the patterns found in records of amplitude scintillations have features with regular structures with a definite period. QP scintillations differ from random scintillations in that random scintillations appear ``noisy'' while QP scintillations closely resemble the sin waveform that is well known from optical interference studies. The ``ringing'' structure of QP scintillations should possess a deep central minimum for events caused by ionospheric irregularities, while QP events with a central maxima are attributable to spurious radio frequency interference from ground transmitters or other satellites. The most widely accepted view on the production of the QP scintillations is that they are caused by reflection from the frontal structure of the irregularities in a manner that has been described previously in looking at the scattering model. This view is supported by the fact that QP scintillations tend to occur for low elevation angles and where the ray path coincides with the direction of the slope of the frontal irregularities. While it is possible that QP scintillations may be caused by irregularities in the E or the F region, the prevalent thinking on the matter suggests that irregularities associated with sporadic-E disturbances are the most likely culprit. The relatively high number of QP scintillations recorded during the day-time, when compared to P type random scintillations, further supports this notion. A final possibility is that QP scintillations may take place following the passage of a TID, either from the frontal structure of the TID or from irregularities produced in its wake.

IV. Ionospheric field-aligned irregularities

Investigations of the small-scale structure of the ionosphere conducted by scintillation studies frequently draw upon information gleaned from sources other than the record of the received amplitude of the transionospheric signal. More traditional ionospheric investigative procedures such as the use of ionosonde data and TEC records are particularly useful in this context in that the known morphology of gross features in the mid-latitude ionosphere can be incorporated into the total body of knowledge available to researchers. This type of information plays a vital role in the formulation and evaluation of theories that describe the manner in which ionospheric disruptions over a range of scale sizes develop. The role played by field-aligned irregularities and their occurrence in conjunctions with large scale disturbances is central to the process of describing scintillation phenomena.

The occurrence of field-aligned irregularities in the F2 region of the mid-latitude ionosphere is well known. These scintillation producing irregularities have especially been found to develop at the same time as range spread-F has been detected by topside and bottom side ionosonde measurements. Early results of comparatives analysis of spread-F studies and scintillation occurrence suggested that the scintillation irregularities were associated with frequency spread-F but this has since been ruled out as a selective bias by the ionosonde technique in finding frequency rather than the range spread-F.

The simultaneous occurrence of scintillations and range spread-F during the night-time ionosphere of the geomagnetic mid-latitude region during solar minimum is thought to explain the relatively high number of scintillations recorded during this time. Measurements in the high and low latitude regions at similar times do not display a concurrent increase in the number of scintillations detected. As was pointed out in the preceding section, the P type scintillations which are the only true mid-latitude scintillations display a similar occurrence enhancement while the S and N type scintillations, belonging to the high and low latitude regions do not. The vertical extent of spread-F in mid-latiude regions, in the order of 60 km, is consistent with the axial ratios determined for the irregularities from scintillations studies of P type scintillations.

An examination of field-aligned irregularities on the cusp of the auroral scintillation oval by Wernik et. al found that the S type scintillations were produced by rod-like inhomogeneities with axial ratios between 5:1:1 and 10:1:1. By contrast, a study by MacDougall (1990) of P type scintillations within the mid-latitude ionosphere established that these ratios may be significantly understated. Theoretical considerations indicate that the weak irregularities thought to be responsible for the scintillation activity may have axial ratios of up to 63:1:1 with the cross-sectional width in the order of 0.8 km. During observations at mid-latitude stations in the United States, results were obtained that indicated preferential elongation of the irregularities in the direction of the magnetic field with mean axial ratios of 44.5:1:1 and minor half-axis widths of 0.3 ± 0.13 km. Speculation that weak, extended, ever-present irregularities aligned with the direction of the magnetic field was raised in the 1960's by such workers as Jones (1969), Parkin (1968) and Singleton and Lynch (1962) during observational campaigns conducted at sites operated by the University of Queensland in the vicinity of Brisbane. These three studies all found that the irregularities responsible for the scintillation activity were due to the presence of field-aligned irregularities in the kilometric range at F region heights, with a marked enhancement of scintillation activity for small aspect angles. Although the studies concluded that the irregularities could be rod-like or sheet-like, this result is likely to be due to the increased sensitivity of the relatively low frequency signals to the presence of sheet-like irregularities. Modern studies operating using higher frequency signals in excess of 100 MHz conclude that the irregularities are rod-like and not sheet-like, as found by topside ionosonde studies that obtain ducted echoes (Dyson 1967).

Similarly, modern studies also differ in their interpretation of the specific ray propagation interactions that produce the scintillations. Earlier studies tended to favor the notion that the interference arose when signals encountered the irregularities at grazing incidence angles and reflection from the surface of the excess electron density distribution with the undeviated portion of the ray were responsible for producing the amplitude scintillation patterns produced. While this theory has not been discounted and is still widely used in qualitative studies, contemporary thinking leans toward the use of diffracting phase screens, be they of the Briggs and Parkin type or power law form. All studies demonstrate a pronounced enhancement in scintillation activity where the ray path coincides with the field point. This aspect angle sensitivity has been noted by many observers, such as Sinno and Minakoshi, although the techniques used to interpret data vary widely. Older studies concentrated on confining the data sets used to variations in the zenith (Briggs and Parkin 1963) or azimuth (Jones 1969) angle while modern studies, using computer calculations to find the aspect angle as a function of zenith and azimuth angle, are not restricted in the same manner.

The manner in which the irregularities form is still a point of contention. The regions of the ionosphere where they form is well described (mainly the F region although some form in the E region) but the specific processes that drive their production are undecided at this time. A satisfactory explanation for the spatial and temporal integrity of the electron density enhancements is not yet fully formed. The most likely candidate at this time would appear to be the generation of differentiated turbulence by the passage of TIDs, probably AGWs moving up from the neutral atmosphere, which halt the diffusion process that is expected to nullify the formation of irregularities beyond 3-4 seconds. Some indicators exist to suggest that this process is aided by electric fields that act in the direction opposite to the gravity forces at work. A magnetic bottling effect that is well known from the theory of plasmas then enhances the concentration of the irregularities in the direction of the Earth's magnetic field. Tomographic analysis of this situation is expected to reveal more information on this topic in the future.

V. Impacts of Ionospheric Scintillation

Ionospheric scintillation affects trans-ionospheric radio signals up to a few GHz in frequency and as such can have detrimental impacts on satellite-based communication and navigation systems (such as GPS-based systems) and also on scientific instruments requiring observations of trans-ionospheric radio signals (eg. radio-astronomy).

Amplitude scintillation directly affects the signal to noise ratio (C/No) of signals in a GPS receiver, as well as the noise levels in code and phase measurements. Amplitude scintillation can be sufficiently severe that the received GPS signal intensity from a given satellite drops below the receivers tracking threshold, causing loss of lock on that satellite, and hence the need to re-acquire the GPS signal(s). This results in reduced accuracy navigation solutions, data loss and cycle slips. The nominal C/No for the L1 signal is about 45dB-Hz, and tracking may be lost when the signal drops below ~25dB-Hz, dependent on the receiver-specific tracking loop.

Since the signal power on the GPS L2 frequency is significantly less than that of L1 (~6dB lower), and civil dual frequency receivers use non-optimal codeless or semi-codeless techniques for tracking L2 which results in lower C/No values, ionospheric scintillation is much more likely to impact the GPS L2 signal.

Phase scintillation, if sufficiently severe, may stress phase-lock loops in GPS receivers resulting in a loss of phase lock. Phase scintillation also has a significant impact on phase-sensitive systems such as space-based radars (eg image defocusing in synthetic aperture radars (van de Kamp et al, 2007)) and some ground-based radio-astronomy facilities (eg SKA/LOFAR prototypes (van Bemmel et al, 2007)).

7.2 Methodology and Measurements

There are numerous measures of ionospheric scintillation. Perhaps the most common of these is the amplitude scintillation index S_4, and the phase scintillation index P_{rms} Ionospheric scintillation models produce statistical measures of the specified scintillation index. To produce maps it is necessary to either specify thresholds of this index or to specify a percentage of time an index is exceeded.

A network of stations setup by various countries across the Arctic and Antarctic regions have been recording data on a continuous basis by launching Polar Scientific Expeditions annually. For the present study we have used the data which was collected by the authors, during the 27th Indian Antarctic Scientific Expedition at Indian base station Maitri from January 2008 to January 2009. Ionospheric scintillation and TEC measurement was performed using the GPS Ionospheric Scintillation and TEC monitor (GISTM) model GSV4004A. The system is NovAtel's dual frequency GPS receiver. Ionospheric total electron content data were recorded with 30 second data sampling in order to reduce processing time. The GPS receiver was set to track GPS signals at one second sampling rate and cut off of elevation angles was set to 40^0.

The GPS observables are biased by the instrumental delay therefore it is necessary to remove these biases for accurate estimation of TEC. The absolute Total Electron Content (VTEC) determination has the capability to remove the instrumental biases both from the receiver and the satellite. The instrument time delay and potential errors are corrected using the code biases.

The purpose of the ISM system is to collect ionospheric scintillation statistics (S_4 and P_{rms}) for all visible GPS satellites (up to eleven satellites) and store these (ISMR) binary data logs on the receiver controller hard disk for post processing. The ISM control software can be programmed to collect the ISMR data logs that are generated every 1 minute. Alternatively, raw phase and amplitude data at 0.02 second temporal resolution (50Hz) and code/carrier divergence at 1s (1Hz) can be recorded from the ISM. These data can be used to reconstruct the statistical scintillation indices, such as S_4 recorded in the ISMR data log, from raw data. This allows the user to modify the parameters used in the derivation of scintillation indices, such as de-trending and filter cut-off parameters.

Amplitude Scintillation: -The amplitude scintillation was monitored by computing the S4 index, which is defined as the standard deviation of the received signal power normalized to the average signal power. It is calculated for each 60 second period based on a 50 Hz sampling rate. Phase scintillation computation is accomplished by monitoring the σ_φ index, the standard deviation of the deterrent carrier phase computed over 1, 3, 10, 30 and 60 s intervals. Although scintillation indices have been widely used to monitor and measure the intensity of scintillation, but their derivation have some problems and errors and hence sometimes become doubtful (Beach, 2006).

The GISM used in this analysis measures both amplitude and phase scintillation. Amplitude scintillation is defined by the S_4 index that is derived from detruded intensities of signals received from satellites. The S_4 index is computed over 60-second intervals and stored in the Ionospheric Scintillation Monitor Receiver (GISM) data log along with the phase measurements. This is referred to as the Total S_4 (or S_{4T}). The normalized S_4 index, including the effects of ambient noise, is defined as follows:

$$S_{4T} = \sqrt{\frac{\langle P^2 \rangle - \langle P \rangle^2}{\langle P \rangle^2}} \quad \dots\dots\dots\dots\dots\dots 1$$

The amplitude measurements are filtered using a Low Pass Filter (LPF) and the effects of ambient noise removed from the S_{4T}. This is achieved by estimating the average signal-to-noise ratio over the 60-second interval. The 60-second estimates are then used to determine the expected S_4 correction (or S_{4N_0}) due to ambient noise. The use of this average signal-to-noise ratio (S/N_0) is feasible because the amplitude scintillation fades do not significantly alter the S/N_0. Knowing the S/N_0, S_{4N_0} due to ambient noise becomes:

$$S_{4N_0} = \sqrt{\frac{100}{S/N_0}\left[1 + \frac{500}{19 S/N_0}\right]} \quad \dots\dots\dots\dots\dots\dots 2$$

Equation 2 is referred to as the S_4 correction (or S_{4N_0}). By subtracting the square of the right hand side of Equation 2 from the square of the right hand side of Equation 1, and replacing the S/N$_0$ with the 60-second estimates. Equation 1 may be modified to give the S_4 index, with the effects of ambient noise removed, as follows:

$$S_4 = \sqrt{\frac{\langle P^2 \rangle - \langle P \rangle^2}{\langle P \rangle^2} - \frac{100}{S/N_0}\left[1 + \frac{500}{19S/N_0}\right]} \quad \ldots\ldots\ldots\ldots 3$$

Phase scintillation: Phase scintillation is quantified by the P_{rms} (or ϕ_{rms}) index which is defined as the standard deviation of the signal phase over a given time interval. This index is measured either in radians or degrees. A P_{rms} greater than ~ 1° is considered to be strong scintillation. At mid-latitudes, P_{rms} rarely exceeds 1° for more than 1% of the time.

Other scintillation parameters: Log of the height-integrated irregularity strength (calculated on LOS paths from the ground to an overhead satellite). A measurements of total power of the electron density irregularities along a vertical path passing through the entire ionosphere.

Scintillation Intensity (SI) index derived from scintillation data recorded on paper chart. Scintillation Intensity index (*SI*) is defined as:

$$SI = \frac{P_{max} - P_{min}}{P_{max} + P_{min}}$$

where P_{max} is the power of the 3rd peak down from the maximum excursion shown during a scintillation occurrence, and P_{min} is the power of the 3rd peak up from the minimum excursion. These values can be readily and rapidly scaled from a calibrated chart. The *SI* index is expressed in decibels (dB). An *SI* value of 15dB corresponds to an S_4 of about 0.6.

7.3. Results and Discussion

In this paper scintillation morphology is described in terms of percentage occurrence in specified threshold level according to intensity and differential phase of S4 – Index respectively when S4 is greater than 0.1, 0.3 and 0.5. The study is divided in hourly, monthly and seasonal variations of S4-Index observed in GPS signals at Maitri Antarctica. Figure 7.4 show the scintillation occurrences with different S4 index values i.e. S4 is greater than 0.1, 0.3 and 0.5. Firstly, when S4 is greater than 0.1, this type of small disturbance is regularly observed most of the time throughout the year during the observation period because polar ionosphere is disturbed most of the time due to solar and magnetic activity disturbances (Base et al., 1988). When S4 is greater than 0.3, the maximum percentage occurrence of 30% is observed in the morning hours (~ 0400 to 0700UT) of winter months and after noon time hours (~1300 to 1800 UT) of summer months. Further when scintillation index S4 is greater than 0.5 maximum occurrences up to 28 % is observed in the morning hours (0500 to 0700 UT) and in the evening hours (1600 to 1800) in winter and summer months respectively.

The hourly scintillation activities for different levels of S4 index during the summer season (November, December, January and February) when solar radiation is present for all the 24 hours in polar region which is shown in Figure 7.5. The percentage occurrences of ~30%, 18% and 12% is observed when S4 is greater than 0.1, 0.3 and 0.5 respectively. In the winter period (May, June, July and August) when solar radiation is completely absent i.e. polar nights, the percentage occurrences of ~18% is observed when S4 is greater than 0.1 whereas at higher s4 it is upto less 10 %. During equinoctial months (March, April, September and October) the occurrence of scintillations is about 19 %, 8% and 5% respectively with S4 is greater than 0.1, 0.3 and 0.5 in the morning as well as in the evening hours.

Similarly in terms of seasonal variability maximum percentage occurrence is observed during winter season as compared to summer season and maximum monthly percentage occurrence of S4 index is observed in September and October month as shown in figure 7.6. From the above observations it is noted that during the period of low solar activity mostly weak scintillations are observed.

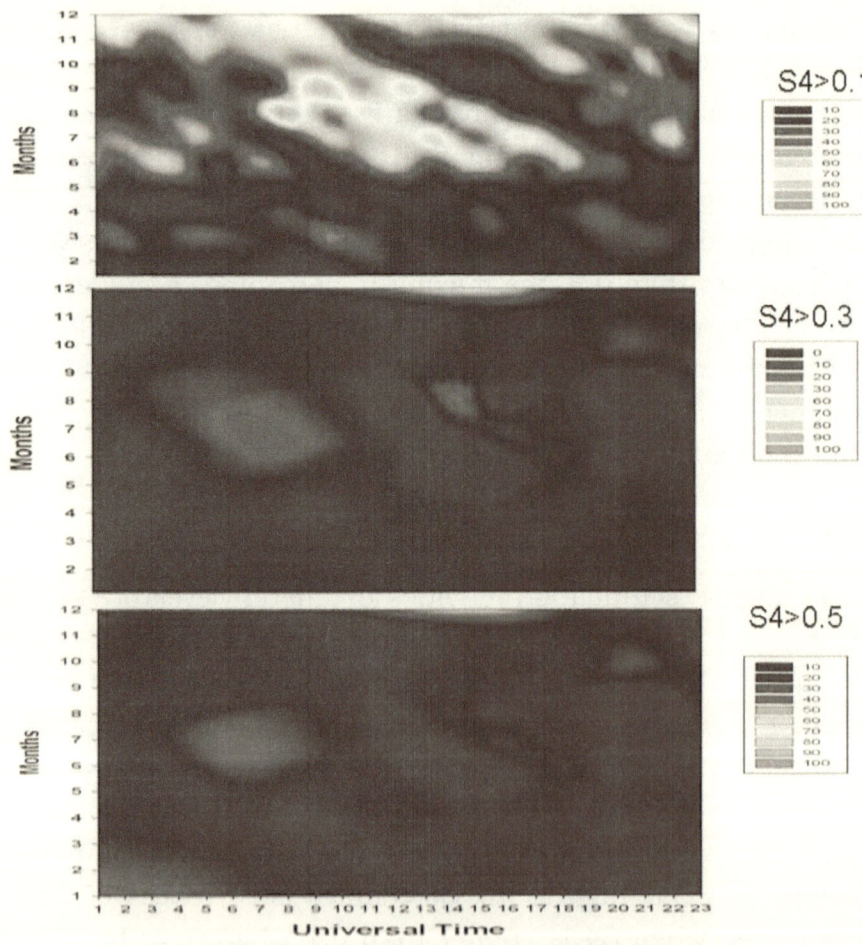

Figure 7.4: Contour plot shows the percentage occurrence at universal time in a year over Maitri, Antarctica.

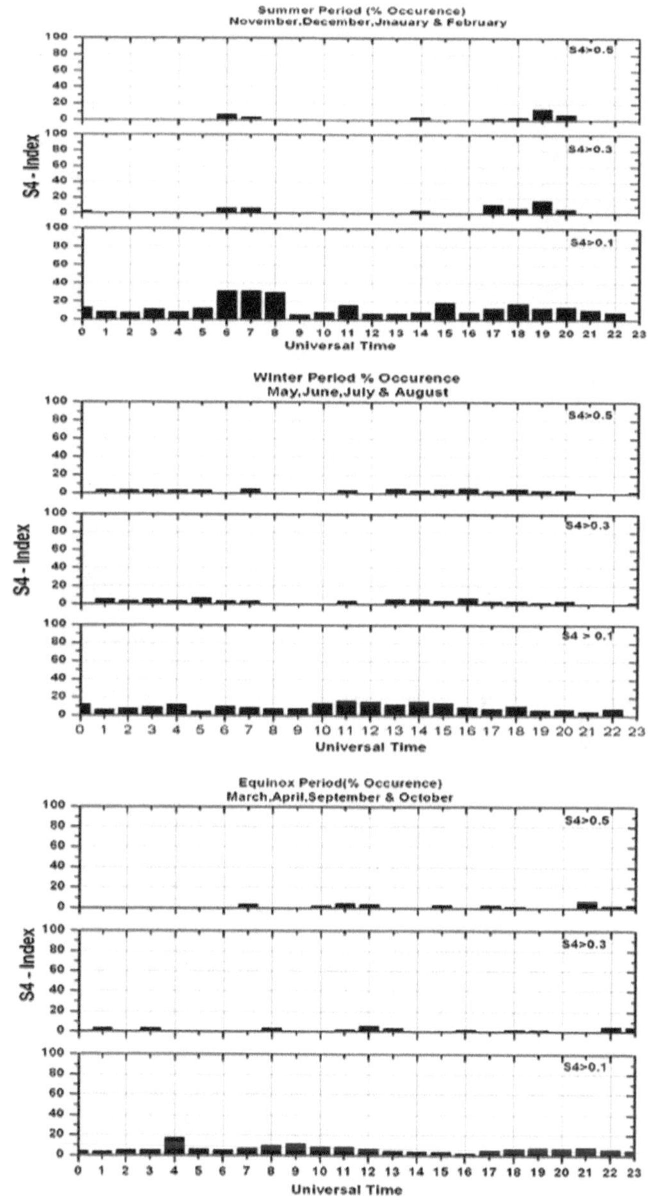

Figure 7.5: Shows the seasonal percentage occurrence of S4 index over Maitri, Antarctica.

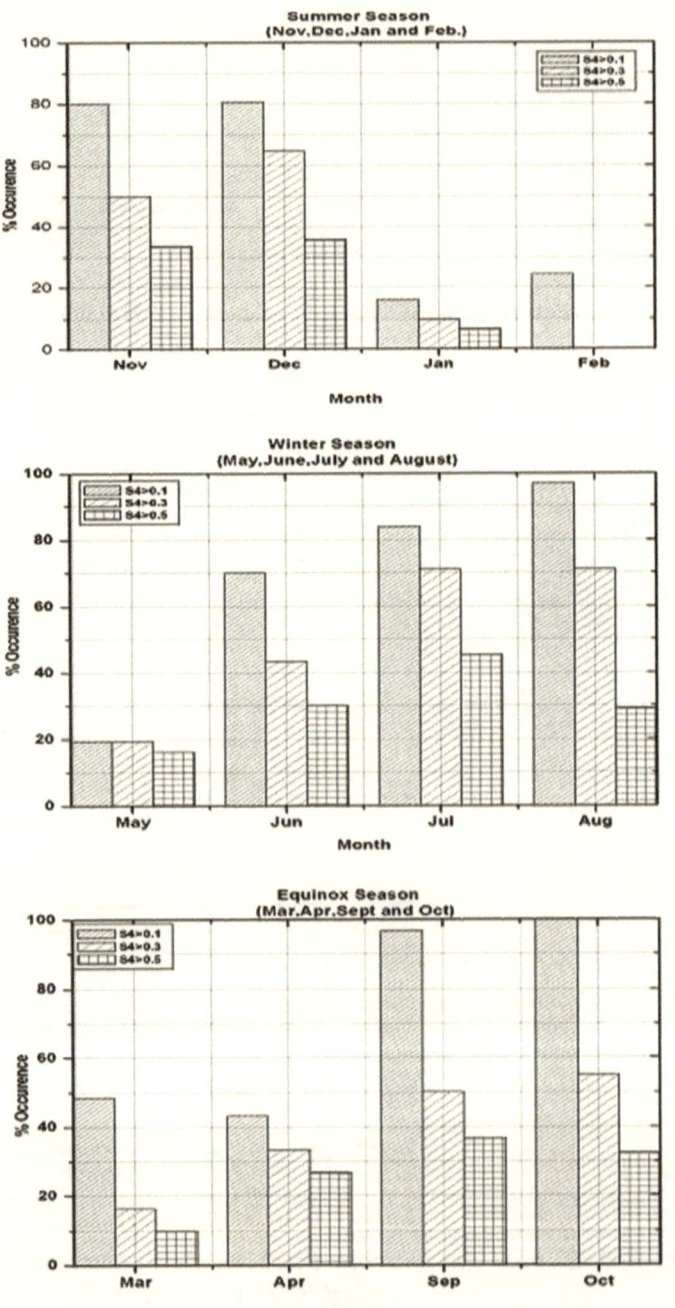

Figure 7.6: Bar plot shows seasonal wise monthly percentage occurrence of S4 index over Maitri, Antarctica.

The study shows that at high latitudes only weak scintillations (S4>0.1) were observed both during day and night time periods. In general at high latitude scintillations occur over night side auroral oval and the polar cap at all local times. The winter polar cap experiences moderate to strong L Band scintillation in association with the so called polar cap patches (Weber et al., 1984). When the interplanetary magnetic field (IMF) is directed southward, patches with high ionization density are observed to enter the polar cap from the day side auroral oval, convected in the anti-sunward direction and eventually exit into the night side auroral oval. One mechanism by which patch formation is achieved corresponds to changing the plasma convection pattern in response to the IMF component during periods of southward Bz (Sojka et al., 1993). During the observation of monthly occurrence of L band scintillation, we have seen the more irregularities occurred in the southern summer months (June to December). High latitudes scintillations are found to be associated with large scale plasma structures known as patches, blobs and sun–aligned arcs (Weber et al., 1984, 1986; Tsunoda, 1988, Basu and Valladares, 1999). High latitude auroral irregularities are formed from a precipitation of energetic electrons along terrestrial magnetic field lines. These electrons are energized through complex interaction between the solar wind and the Earth's magnetic field, resulting in optical and UV emissions commonly known as the auroras. Evidences indicate that the dayside auroral oval plays a major role in the formation of large scale ionization structures in the polar ionosphere (Weber et al., 1984). These structures convect across the polar cap and cause destabilization of the plasma and then develop intermediate scale irregularities by the action of the gradient drift instability mechanism (Tsunoda, 1988). The destabilization process also includes the current convective and Kelvin-Helmholtz instability (Basu et al., 1986). It is established that precipitation of soft particles into the F region may play a direct role in the irregularity formation (Basu et al., 1983; Kersley et al., 1988). In addition to patches, the sheared electric field in the cusp region is also a source of localized intermediate scale irregularities (Basu et al., 1988). Since patches are associated with high plasma density, scintillations of satellite signals due to irregularities in the velocity shear region are expected to be weaker than patch induced scintillations (Basu et al., 1988).

7.4. Conclusion

The study can be summarized as follows. The weak scintillations (S4>0.1) are observed all the 24 hour of the day in almost all the seasons whereas during morning and afternoon hours slightly higher magnitude scintillations (S4 < 0.5) are also observed during the solar minimum period of 2008. Season wise maximum occurrence is noted during summer months whereas in a winter and equinox seasons scintillations are observed mostly in early morning hours as well as night hours. As regarding month wise maximum occurrence is observed between June and December months Present work further confirm the earlier observations reported by various other workers.

CHAPTER VIII

SOLAR CYCLE VARIATION OF CRITICAL FREQUENCY FOF2 FOR SOLAR CYCLE 23 OVER SYOWA, ANTARCTICA

8.1. Introduction

The sun follows a periodic cycle of activity, this cycle, called solar cycle, and is the periodic recurrence of sunspot or darker relatively cool regions at the sun's surface. During the solar cycle the sun emits a wide verity of solar radiation originating in different part of the solar atmosphere solar ultraviolet (UV) irradiance (115-420 nm) plays a dominant role in the temperature distribution, photo chemistry, and overall momentum balance in the stratosphere, mesosphere and lower thermosphere (Ozgue et., al, 2007) However, Electron concentration in the F2-region of the ionosphere is primarily due to ionization of the neutral atmosphere by the solar UV radiations. These radiations are now known to show very definitive solar cycle variations. Consequently, electron concentrations and thus, the critical frequency of the F2-region (foF2) is also expected to reflect these variations. Although there were no solar UV measurements during the early years of ionospheric research, traditionally, the smoothed monthly mean sunspot number (R12) was considered a primary index of solar activity for prediction of ionospheric parameters. The dependence of foF2 on R is "poisoned" by the phenomenon of hysteresis, which has been known for a long time (Naismith and Smith, 1961; Naismith et al., 1961; Huang, 1963; Rao and Rao, 1969; Muggleton, 1969). However, several efforts have been made to introduce a new solar activity index for this prediction. Lakshmi et al. (1998) proposed to use EUV data for the long-term predictions of the monthly median ionospheric parameters. Kane (1992) reported that in the absence of solar EUV data, solar radio flux at 10.7 cm may be better than sunspot numbers when making ionospheric predictions.

Kleczek (1952) introduced the quantity 'FI = it' to quantify the daily flare activity over 24 hours per day. He assumed that this relationship roughly gave the total energy emitted by the flare and named it 'flare index' (FI). In this relation, it represents the intensity scale of importance and 't' the duration of the flare in minutes. Catalogues of flare activity using Kleczek's method are given for each day from 1936 to 1986 by Kleczek (1952), Knoska and Letfus (unpublished), Knoska and Petrasek (1984), Atac, (1987) and for 1986–1995 by Atac, and Ozguc (1998). The flare index is an interesting parameter and is of value as a measure of the shortlived activity on the Sun (Atac and Ozgue et., al. 2000). Some powerful and explosive events on the sun, such as coronal mass ejections (CMEs) can lead to worldwide disturbances. These events can, and do, have an impact on the performance and reliability of space and ground based operational system (Lambour et., al. 2003, Atac and Ozguc et., al. 2005).

Solar activity variations are manifesting themselves not only in electromagnetic radiation from radio frequencies of a few kHz to powerful gamma rays but also in particle flux. In broad physical terms, solar activity may be understood in terms of the properties and the behavior of the magnetized solar plasma. Solar structures and phenomena all arise from magnetic fields embedded in dynamic plasma. Various forms appear at all latitudes from the poles to the equator and at all space scales from several hundred to many hundred thousand km. Some of these structures are remarkably long-lived, and a variety of structures are observed over the entire electromagnetic spectrum from radio, through visible to X-rays and gamma rays. Continuous observations are required to catch short-lived but infrequent phenomena. One important example is solar flares. Images of the Sun show that solar flares are one of the most powerful and explosive of all forms of solar activity. Many studies in the solar-terrestrial field classified solar flares as one of the most important solar events affecting the Earth (Atac et., al. 2000)

Antarctica has an influential role to play in research into long-term change processes. It provides important opportunities for testing critical hypotheses relating observations to current theories, both because of its advantageously unique geophysical attributes and because of its remote location with respect to localized anthropogenic sources. One of the goals of this study was to demonstrate, which solar activity is more convenient for ionospheric predictions.

Ionospheric variations can be considered in time scales of (I) Day to Day variability (II) semiannual (III) Annual, and (IV) solar Cycle. In this present study, examination is made only of the long – term (Solar – Cycle) variations of the solar activity indices and ionospheric parameter foF2. we investigate the response of the ionosphere to the solar activity by using the flare index, the sunspot number (Rz), the solar radio flux at 10.7 cm. and coronal mass ejections (CMEs) occurrence during the 23rd solar cycle over the Japanese Antarctic Station Syowa (Geographic - 69°. 0'S and 14.79 E, Geomagnetic 39.34 S and 55.90 E).

8.2 Ionospheric Variations

The ionosphere is not a stable medium that allows the use of the same frequency throughout the year, or even over 24 hours. The ionosphere varies with the solar cycle, the seasons and during the day.

8.2.1. Daily Variations

Frequencies are normally higher during the day and lower at night (Figure 8.1). After dawn, solar radiation causes electrons to be produced in the ionosphere and frequencies increase rapidly to a maximum around noon. During the afternoon, frequencies begin falling due to electron loss and with darkness the D, E and F1 regions disappear. Communication during the night is by the F2 (or just F) region only and attenuation is very low. Through the night, a maximum frequency gradually decreases and reaches to their minimum value just before dawn.

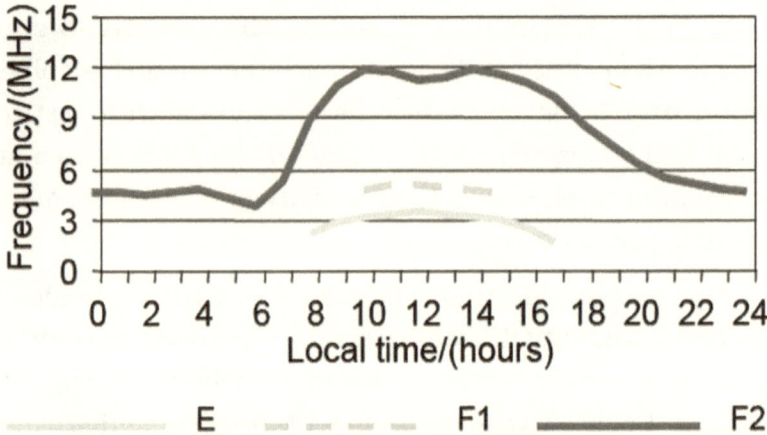

Figure 8.1: E, F1 and F2 layer maximum frequencies throughout the day.

8.2.2. Seasonal Variations

E region frequencies are greater in summer than winter. However, the variation in F region frequencies is more complicated. In both hemispheres, F region noon frequencies generally peak around the equinoxes (March and September). Around solar minimum the summer noon frequencies are, as expected, generally greater than those in winter, but around solar maximum winter frequencies tend to be higher than those in summer. In addition, frequencies around the equinoxes (March and September) are higher than those in summer or winter for both solar maximum and minimum. The observation of winter frequencies often being greater than those in summer is called the seasonal anomaly.

8.2.3. Variations with Latitude

Figure 8.2 shows the variations in the E and F region maximum frequencies at mid-day (Day hemisphere) and mid-night (Night hemisphere) from the pole to the equator. During the day, with increasing latitude, the solar radiation strikes the atmosphere more obliquely, so the intensity of radiation and the daily production of free electrons decrease with increasing latitude. In the F region this latitude variation persists throughout the night due to the action of upper atmospheric wind currents from day-lit to night-side hemispheres (see for example, IPS HF Radio Propagation Course and Manual - http://www.ips.gov. au/Products_and_Services/2/2).

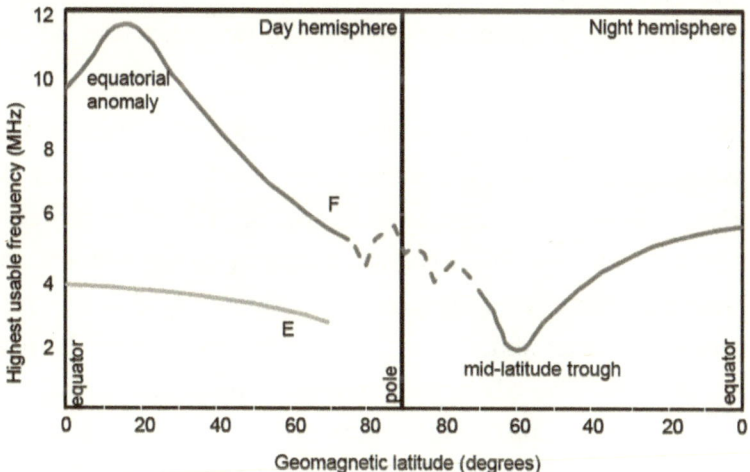

Figure 8.2: Ionospheric Latitudinal variations.

Deviations from the general low to high latitude decrease are also apparent. Daytime F region frequencies peak not at the geomagnetic equator, but 15 to 20° north and south of it. This is called the equatorial anomaly. Also, at night, frequencies reach a minimum around 60° latitude north and south of the geomagnetic equator. This is called the mid-latitude trough. Communicators, who require communications near the equator during the day and night around 60 ° latitude at night, should be aware of these characteristics. The frequencies can change with latitude near the mid-latitude trough and equatorial anomaly, so a variation in the reflection point near these by a few degrees may lead to a large variation in the frequency supported.

8.2.4. Variations due to the Solar Cycle

The Sun goes through a periodic rise and fall in activity which affects HF communications; solar cycles vary in length from 9 to 14 years. At solar minimum, only the lower frequencies of the HF band will be supported (reflected) by the ionosphere, while at solar maximum the higher frequencies will successfully propagate (Figure 8.3). This is because there is more radiation being emitted from the Sun at solar maximum, producing more electrons in the ionosphere which allows the use of higher frequencies.

There are other consequences of the solar cycle. Around solar maximum there is a greater likelihood of large solar flares occurring. Flares are huge explosions on the Sun which emit radiation that ionizes the D region, causing increased absorption of HF waves. Since the D region is present only during the day, only those communication paths which pass through daylight will be affected. The absorption of HF waves travelling via the ionosphere after a flare has occurred is called a short wave fade-out. Fade-outs occur instantaneously and affect lower frequencies the most. If it is suspected or confirmed that a fade-out has occurred, it is advisable to try using a higher frequency. The duration of fade-outs can vary between about 10 minutes to several hours, depending on the duration and intensity of the flare.

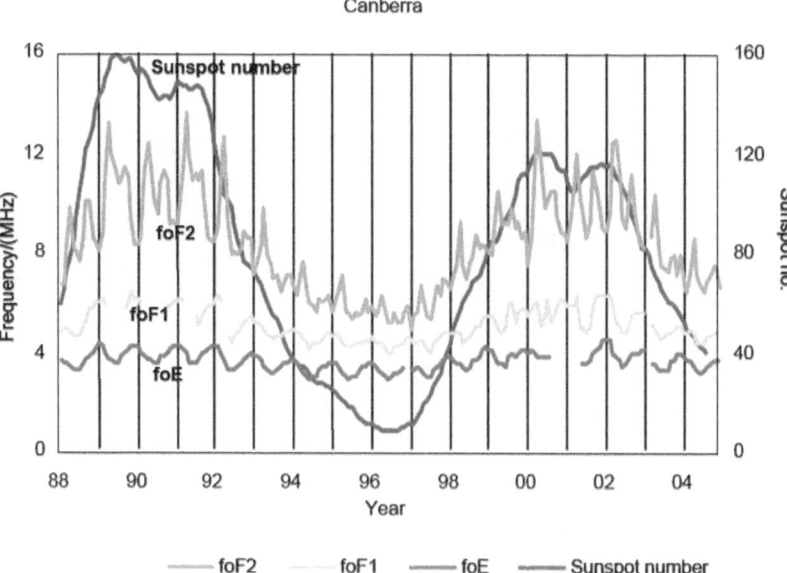

Figure 8.3: The relationship between solar cycles and maximum frequencies supported by E, F1 and F2 regions. Vertical lines indicate the start of each year. Note also the seasonal variations.

8.2.5. Solar cycle

The solar cycle (or solar magnetic activity periods) has a period of about 11 years. The cycle is observed by counting the frequency and placement of sunspots visible on the Sun. Solar variation causes changes in space weather and to some degree weather and climate on Earth. It causes a periodic change in the amount of irradiation from the Sun that is experienced on Earth. This also increases solar minimums, while the solar cycle takes place the plasma currents pulls sunspots inside the sun & the sunspots are reborn. It is one component of solar variation, the other being a periodic fluctuation.

Powered by a hydro magnetic dynamo process, driven by the inductive action of internal solar flows, the solar cycle:

✦ Structures the Sun's atmosphere, its corona and the wind;

✦ Modulates the solar irradiance;

✦ Modulates the flux of short-wavelength solar radiation, from ultraviolet to X-ray;

✦ Modulates the occurrence frequency of solar flares, coronal mass ejections, and other geoeffective solar eruptive phenomena;

✦ Indirectly modulates the flux of high-energy galactic cosmic rays entering the solar system.

8.2.6. Solar Indices and its variation with solar cycle

Solar variation is the change in the amount of radiation emitted by the Sun and in its spectral distribution over years to millennia. These variations have periodic components, the main one being the approximately 11-year solar cycle (or sunspot cycle). The changes also have periodic fluctuations. In recent decades, solar activity has been measured by satellites, while before it was estimated using 'proxy' variables. Scientists studying climate change are interested in understanding the effects of variations in the total and spectral solar irradiance on Earth and its climate.

8.2.6.1. Sunspot Numbers

The abundance of sunspots on the sun varies on timescales from a few hours to many years. Historically, an index called the 'sunspot number' has been used to quantify the abundance of spots. This index is still in wide use today, although for some purposes it has been replaced by more readily and consistently measures indices such as the 10 centimeter solar flux. The main advantage of the sunspot number is that it is the only index for which we have a long and detailed historical record.Sunspot Number (here denoted R) is defined as:

$$R = K (10G + I)$$

where G is the number of sunspot groups visible on the sun; I is the total number of individual spots visible; and K is an instrumental factor to take into account differences between observers and observatories.

Sunspot Number as an index can be defined on a daily basis but because of the large day-to-day variation is usually averaged over longer periods, the most common being the monthly and the yearly average. When averaged over a year, the sunspot number varies smoothly charting the progress of the solar cycle (Figure 8.4). On the other hand the daily and the monthly averages exhibit considerable variation with respect to the yearly curve. This variation is due to bursts of rapid solar region growth often associated with solar flares and other interesting events.

The most widely quoted average sunspot number is the Zurich number (Rz) which was replaced from January 1981 with the International Sunspot Number (RI). The American Sunspot Number is another series to which the IPS Culgoora Observatory contributes its observations.

A daily count of the number of sunspots visible on the Sun (the Wolf number or Zurich sunspot number) shows a periodic variation with maxima occurring (on average) every 11 years. During a solar sunspot minimum, there may be no spots visible on the Sun for several days, and the number present during maxima varies with each cycle.

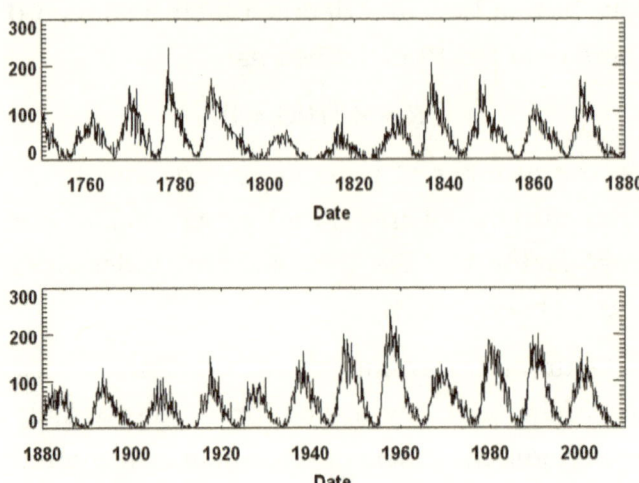

Figure 8.4: Sunspot cycle variations

8.2.6.2. Radio Flux (F10.7)

Sun emits radio energy with a slowly varying intensity. This radio flux, originating from atmospheric layers high in the Sun's chromosphere and low in its corona, changes daily with the sunspot number, and thus also follows the solar cycle. The Figure 8.5 displaying the flux at 2800 MHz or 10.7 cm wavelength over the entire solar disk (in here, monthly averages, although daily values exist too). This value is called F10.7, and it is available from continuous routine measurements.

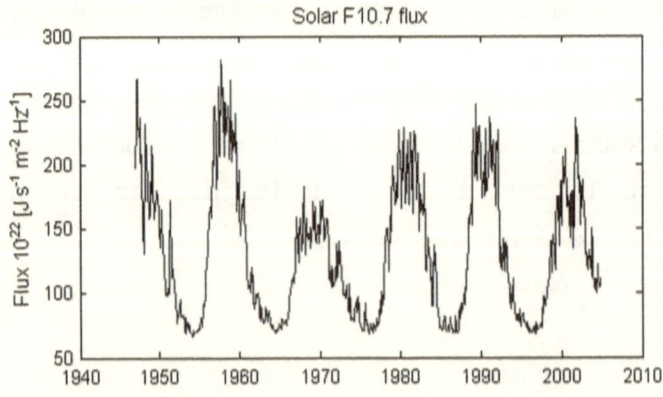

Figure 8.5: Radio Flux index variation

8.2.6.3. Solar Flare Index (FI)

The Solar Flux Index (SFI) is an estimate of solar particles and magnetic fields reaching our atmosphere. Higher numbers mean more solar wind is reaching the earth thus better propagation. The daily sums of the index for the northern and the southern hemispheres and for the total surface are divided by the total time of observation of the day (Figure 8.6). Because the time coverage of flare observations is not always complete during the day (sometimes 75% or 90%), it is corrected by dividing by the total time of observations of that day to place the daily sum of the flare index on a common 24- hour period (Atac and Ozguc 1998)

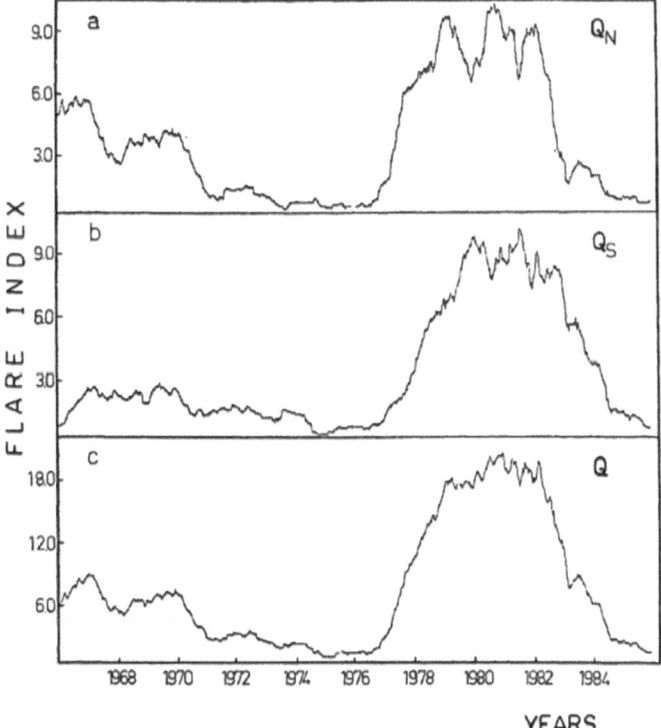

Figure 8.6: Time history of 365-day moving average of the daily value of the flare index for the period 1966-1986, top, middle and bottom curve indicate the flare index value of northern hemisphere, southern hemisphere and whole disk, respectively, for the some period.

8.2.6.4. Corona Mass ejections (CMEs)

Some of the most dramatic space weather effects occur in association with eruptions of material from the solar atmosphere into interplanetary space. These eruptions are known as coronal mass ejections, or CMEs. A large CME can contain 10.0E16 grams (a billion tons) of matter that can be accelerated to several million miles per hour in a spectacular explosion. Solar material streaks out through the interplanetary medium, impacting any planets or spacecraft in its path. The coronal image below shows the release of a CME at the Sun.

The event occurring here is on the side of the Sun – or the limb – which means that it will not affect us here on Earth. Sometimes, however, CMEs occur on the front side of the Sun in a location directly in front of Earth. These events appear to be very different when viewed from Earth. Instead of looking like a "bubble" of plasma, they form a circle of bright light around the Sun. This light is much dimmer than the Sun itself which is why you need to put a disk in front of the disk of the Sun in order to see what goes on around it.

Near solar activity maximum, the sun produces about 3 CMEs every day, whereas near solar minimum it produces only about 1 CME every 5 days. The faster CMEs have outward speeds of up to 2000 kilometers per second, considerably greater than the normal solar wind speeds of about 400 kilometers per second. These produce large shock waves in the solar wind as they plow through it.

CMEs are sometimes associated with short periods of explosive energy release, known as solar flares. These flares frequently occur in active regions during the period around solar maximum. An example of a flare associated with an Earthward-directed CME is shown below. Flares have lifetimes ranging from hours for large gradual events down to tens of seconds for the most impulsive events. During a very strong flare, the solar ultraviolet and x-ray emissions can increase by as much as 100 times above even active-region levels. During solar maximum, approximately one such flare is observed every week. Flares heat the solar gas to tens of millions of degrees. The heated gas then radiates strongly across the whole electromagnetic spectrum from radio to gamma rays. The largest of these explosions are so bright that they can even be seen from Earth in visible light. The picture below shows the flare associated with an Earthward-directed CME event which occurred on May 12, 1997. The January 6, 1997 CME shown above did not have a flare associated with it.

Flares can accelerate protons and electrons that travel to Earth directly from the Sun along the interplanetary magnetic field (which "channels" the charged particles). These contribute to the high-energy particle environment in the vicinity of the magnetosphere if Earth's location is magnetically connected to the flaring region by the interplanetary magnetic field.

Major flares can be accompanied by energetic protons, which can reach Earth within 30 minutes of the flare's peak. During such an event, Earth is showered with highly energetic solar protons released from the flare site. Some of these particles spiral down Earth's magnetic field lines, reaching the upper layers of our atmosphere. These particles show up as tiny white spots in the images taken by spacecraft cameras.

The area between the Sun and the planets is called the interplanetary medium. It is often described as a vacuum, but this is not true. It is actually a turbulent area dominated by the solar wind, which flows at velocities of approximately 250-1000 km/s (about 600,000 to 2,000,000 miles per hour). Other characteristics of the solar wind (density, composition, and magnetic field strength, among others) vary with changing conditions on the Sun. In general, disturbances in the solar wind arrive at Earth 2-4 days after leaving the Sun - the CME on January 6-7, 1997 did not arrive at Earth until January 10, 1997. This CME belongs to a particular subset of CMEs, termed magnetic clouds, which usually have a greater effect on the Earth. The interplanetary space signature of a magnetic cloud is very distinct. The most easily recognized characteristics are strong magnetic fields and a large and smooth rotation of the magnetic field direction.

When these disturbances arrive at Earth, they do not always have the same effect. The factor in determining how much the Earth will be affected by a CME is the direction of the magnetic field – in particular, the north-south direction, or 'z' component. When the z component is positive, this corresponds to a northward field, which has little or no effect on the Earth. When the z component is negative, however, this corresponds to a southward field. When the interplanetary magnetic field is southward, it opposes the direction of the Earth's magnetic field. In the same way that the different poles of a bar magnet attract (in contrast to like poles repelling), an interaction between the two magnetic fields will occur, allowing the energy from the solar wind to enter the Earth's protective shield – the magnetosphere. It is clear from the above plot of the interplanetary magnetic field for January 9-11, 1997, that the z-component was first negative, southward, and slowly turned northward throughout the event. This means that it had an effect on the Earth – it was "geoeffective".

These solar wind disturbances can trigger global changes in Earth's magnetic field and particle populations, called magnetic storms. A magnetic storm is a period when the magnetic field measured on Earth is highly disturbed and auroras are produced. They generally last several hours to several days. The Dst – Disturbed Storm Time – Index is a measure of the magnetic field measured at Earth.

8.2.6.5. Coronal mass ejection Index

The number of CMEs flare eject from the sun which is measured by their intensity is called CMEs index, some time its called coronal index introduced by Rybansky in 1957, he suggested The coronal index (CI) of solar activity is the irradiance of the Sun as a star in the coronal green line (Fe XIV, 530.3 nm or 5303 A). It is derived from ground-based observations of the green corona made by the network of coronal stations.

8.3. Data and Methods

The F2 layer critical frequency (foF2) is one of the most important parameter in the ionosphere which is observed regularly by several observatories and it is allow to us to examine the relation of F2 layer with the solar activity indices. In this present study we are using the hourly data of foF2 which is taken from Japanese scientific research station Syowa, Antarctica, available on National Institute of Information and Communications Technology (NICT), Japan.(http://wdc.nict.go.jp/iono/hp2009/ISDJ/index-E.html).

The daily data of international sunspot number provided by the Sunspot index Data Center of Royal Observatory of Belgiam were used for our analysis. These data represent the definitive relative numbers of the sunspots calculated on the basis of all observations available from different observatories.

The flare index is an interesting parameter and is of value as a measure of the short-lived activity on the sun. In this study the results of the determination of the flare index for solar cycle 23rd are presented. This data collected from National Geophysical Data center (NGDC) website.

Coronal mass ejections (CMEs) are also powerful parameter for responsible of solar activities, here we are using occurrence of CMEs during the 23rd solar cycle which is available on National Geophysical Data center (NGDC) website.

8.4. Results

Figure 8.7, shows the plots of the 12 month moving average of solar indices and foF2. we are presenting individual behavior of single parameters during the time period between 1996 – 2008 which covers almost 23 solar cycle. All these indices show a monotonic increase from the beginning of the solar cycle to their maxima. The following may be noted:

1. The critical frequency of F2 layer observed maximum during the peak period of solar cycle 2001 – 2002, but during the month from May, Jun, July there is no significant variation observed.

2. Maximum peak of smooth sunspot number (Rz) observed during the maxima phase of solar cycle between May to October months.

3. Radio Flux (F10.7) also shows maxima during the maximum phase of solar cycle but it increases during the August to December months.

4. Flare Index shows the maximum peak in 2000 with solar cycle period and between Jun to August month.

5. The occurrences of CMEs were observed maximum between the years 2000 to 2002 that was maximum phase of 23rd solar cycle. The maximum occurrence of CMEs was noticed in November month.

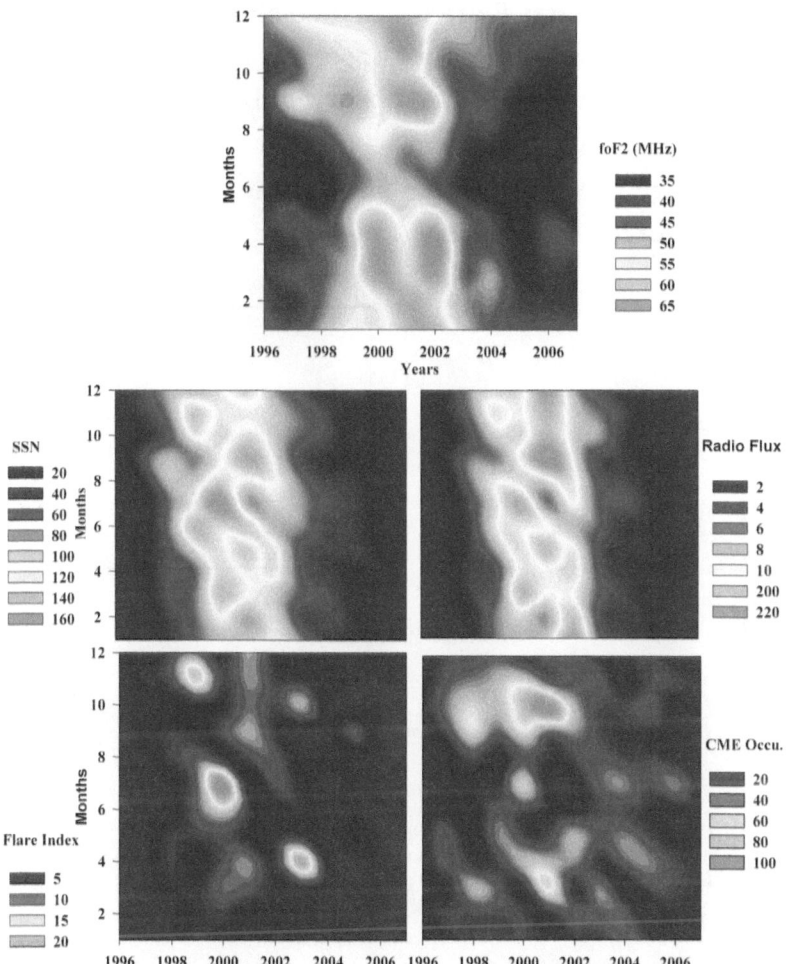

Figure 8.7: behavior of critical frequency foF2 and solar indices during the solar cycle 23 over Syowa, Antarctica

Figure 8.8, shows the individual variation with foF2, the following observation noticed from this figure.

1. Sunspot number (Rz) shows two maxima during the solar cycle.

2. Radio Flux (F10.7) shows the tow maxima with one peak of foF2 another interesting result was observed that second peak was greater than first peak.

3. Solar parameter Flare Index also shows two peaks during the solar cycle first was observed on 48 month of solar cycle and second was observed between on 84 to 96 month of the solar cycle.

4. The occurrence of CMEs shows the single peak variation with fof2 during the solar cycle.

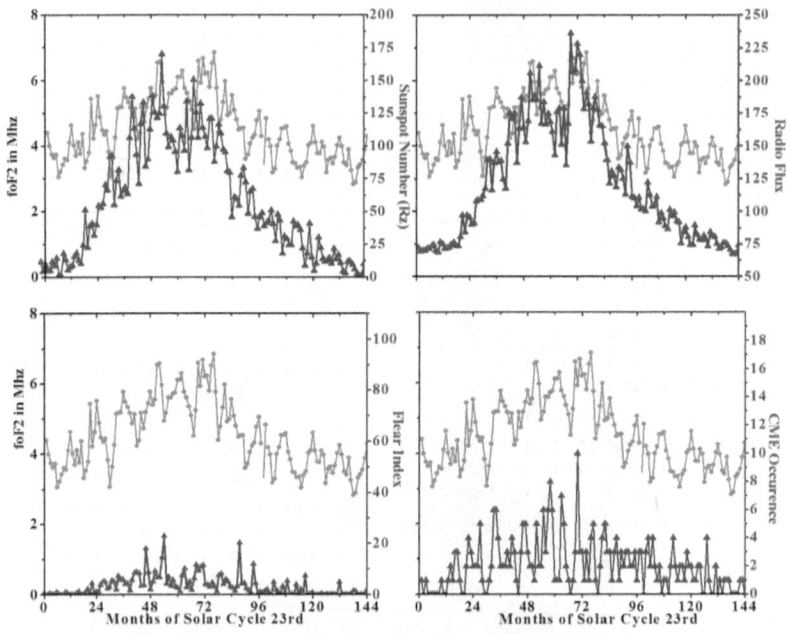

Figure 8.8: Variation of solar indices with critical frequency FOF2

Figure 8.9, shows the peak to peak variation between foF2 and solar indices, in the figure first shows the peak to peak variation between sunspot numbers (Rz) and foF2, it is clear from figure sunspot and foF2 both are equally fluctuate with solar cycle and peak observed during the maxima of solar cycle. In the second figure shows the variation between radio flux (F10.7) and foF2, this variation shows radio flux vary with foF2 during the solar cycle. In the third graph shows the variation between flare index (FI) and foF2, this is showing flare index also vary with solar cycle and it increases with foF2 and decreases with solar activity. The last figure's defining the variation between occurrence of CMEs and foF2 and its peak was observed during the maxima of solar cycle.

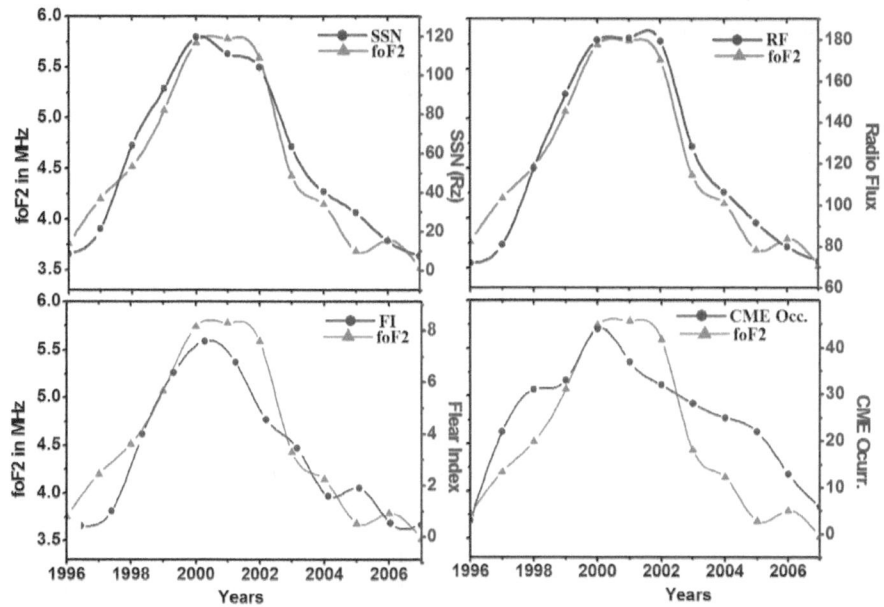

Figure 8.9: Peak to peak variation between solar indices and critical frequency FOF2 during the solar cycle 23, Syowa Station, Anatarctica.

In order to determine the relationship between foF2 and the solar activity parameters a single regression analysis was carried out for Syowa research station, Antarctica, and results are presenting the relationship is linear. This is demonstrated in the figure 8.10 where the variations of foF2 with the solar parameters are shown for the solar cycle 23. There figures contain the observed data and the regression fit, correlation coefficients and the relations between foF2 and the solar parameters.

Figure 8.10, display the variations in the curves of hysteresis between foF2 and the solar activity indices for the 23 solar cycles. We my noted that the sunspot and foF2 hysteresis shows generally lower foF2 for rising branch as compared with falling branches of solar cycle. The second graph shows the single regrration between foF2 and CMEs occurrence, this hysteresis shows the strong correlation during the rising phase as compare to falling phase. In the hysteresis of foF2 and flare index, shows in strong correlation during the rising phase as compare to decreasing phase of the solar cycle. In the last hysteresis between foF2 and radio flux shows the equal correlation in the both phases.

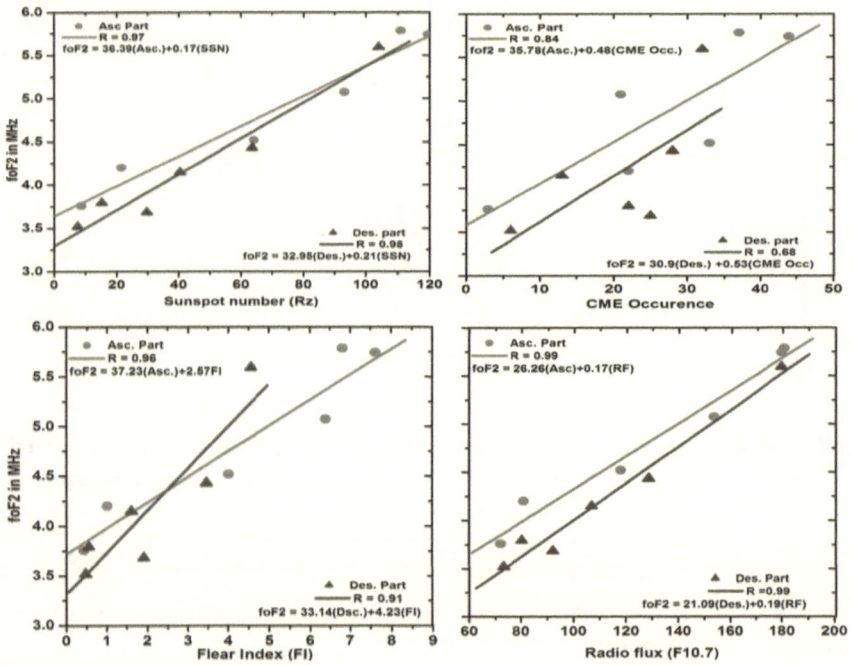

Figure 8.10: Twelve months moving average of foF2 over Syowa station, Antarctica, versus solar indices. Full dots refer to the ascending phase (1996 – 2000) of solar cycle 23 and triangles the descending phase (2001 – 2008). The single regression fits are shown as solid lines for the both the phases of solar cycle 23.

8.5. Discussion

Being able to express aspects of solar activity by many indices, such as the sunspot number, the 2800MHz radio flux, flare index, CMEs occurrence, etc., are useful for studying the Sun's long-term behavior and its interaction with our near Earth space environment. Long-term predictions of the critical frequency have traditionally been based on the relationship between the predicted ionospheric parameters and 12-month running mean of the sunspot number (Rz). The dependence of foF2 on Rz (or R12) is "resentful" by the phenomenon of hysteresis, which has been known for a long time. For a given station and a constant value of the solar activity indices, foF2 differs for the ascending and the descending parts of the 11-year solar cycle. In this connection, we investigated the solar cycle variation of foF2 by using the solar flare index and the CMEs occurrence, relative sunspot number, and the 2800MHz radio flux. The examination of the correlation between the monthly average values of foF2 and several indices of solar activity showed that independently from the kind of index and location, significant hysteresis is present during cycles 23. The explanation for the lag is in the dual peak structure of indices. The first peak is related to sunspots' CME activity and the second peak is thought to be caused by fast solar wind streams, which increase during the declining phase of the solar cycle as more and more mid-latitude coronal holes appear on the solar surface (Legrand & Simon 1985). Indeed, according to Abramenko et al. (2010), the declining phase of the 23rd solar cycle displayed an excess of low-latitude coronal holes. While our analysis supports the Echer et al. (2004) report, the CME speed index, newly introduced, in this study may refine the explanation for the cause of the second peak in the geomagnetic activity.

Although the phenomenon of the ionospheric hysteresis between foF2 and Rz has been known for a long time, a linear relationship between these two parameters is used in forecasting and in long term trend estimations. According to our findings the hysteresis magnitude varies non-systematically with the solar cycles, so the conclusion of the hysteresis into the long-term ionospheric predictions seems not suitable.

The linear correlation between the solar activity indices and foF2 is very strong during the ascending and descending branches of the two cycles (Figure 8.10). The slope of their linear fits shows variations from cycle to cycle, as well as index to index. Solar ionizing flux, meteorological influences, and solar wind conditions are the origins of changes in state of the ionosphere. All of these effects are dependent on local time, latitude, and season (Forbes et al., 2000). To avoid the seasonal variation we applied a 12-month running average to the time series of solar activity indices and foF2. In spite of this, in Figure 8.10, one can see that the hysteresis magnitude is high in different solar cycles depending on different ionospheric stations. So, we conclude that this phenomenon is due to the latitude and meteorological influences as well as solar wind conditions. This conclusion is supported by the results of Forbes et al. (2000).The hysteresis effect in ionospheric parameters, such as foF2, may be compatible with a geomagnetic control for each solar cycle. Geomagnetic disturbances are accompanied by large changes in the ionospheric F2 layer. Although the ionospheric response to geomagnetic activity is highly complex due to the many physical processes involved, there are underlying trends that are useful in characterizing the ionosphere response to storms (Fuller-Rowell et al., 2000; Araujo-Pradere et al., 2002; Rishbeth and Field, 1997; Field and Rishbeth, 1997). Taking into account that geomagnetic activity is higher on average during the descending phase of the solar cycle than during the ascending phase, a clockwise or counter-clockwise hysteresis should be expected at a location depending on its prevalent negative or positive ionospheric storms. This implies negative or positive hysteresis magnitude, respectively.

Ozguc et. al., (1998) showed that to use flare index than any other solar index may be more adequate. They found this result only for solar cycle 21. However, the results for cycles 22 and 23 do not support this conclusion.

In a very recent study (Kane 2006) showed that most of the solar parameters as well as the (foF2) show two maxima, with the second maximum higher than the first maximum. He also found that the magnitudes of the second maxima relative to the first one were different for different solar indices. We found the same result with our parameters.

Therefore, we may add the flare index and CMEs occurrence at polar stations (Syowa, Antarctica) results, since the data set of that study does not contain the flare index and those stations. Ionospheric foF2 changes differ from location to location, indicating that direct linear relationship with solar intensities is not maintained, and complex effects of other parameters are involved (Kane, 2006). Our results support this conclusion. Many previous researches, which were done with the ascending branch data, have resulted in an extensive range of the solar periodicities. This is not easy to explain and indicates that the problem of solar periodicities is still awaiting more systematic efforts. Hence over an 11-year solar cycle the amplification sometimes regenerates more polar field and sometimes less. Hathaway et al. (2003) have reported strong observational evidence that the speed of deep meridional flow toward the equator is driving the sunspot cycle. Obviously, other mechanisms, such as fluctuations in the meridional flow (Hathaway, 1996), believed to be a product of turbulent convection and variations in the gradient of the rotation rate, and also contributed to the cycle amplitude variations. The differences in the speed of the meridional circulation during cycles with different amplitude and all the mechanisms mentioned above can act as an intrinsic dynamics which would explain the midrange solar activity periodicities.

8.6. Conclusion

This work investigated the dependence of foF2 on the solar activity. Four solar activity indices namely flare index, relative sunspot number, solar flux at 10.7 cm, and CMEs occurrence indexes as well as Syowa, Antarctica foF2 monthly average data are used. These indices and foF2 data provide a good opportunity to study the solar activity variability in the ionosphere. The conclusions can be drawn as follows:

The individual variation of critical frequency foF2 and solar indices (Figure 8.7) demonstrate the dependency of each other during the solar cycle but this variation was different during the months, which is depend on solar activity and polar ionospheric behavior.

The variation of solar indices with critical frequency foF2 (Figure 8.8) shows simultaneous with each other during the rising and falling phases of solar cycle.

The peak to peak variation between monthly average of critical frequency foF2 and solar indices parameter (Figure 8.9) is evidence for the absolute dependency for each other.

The linear correlation between the solar activity indices and foF2 is very strong during the ascending and descending branches of the two cycles (Figure 8.10). The slope of their linear fits shows variations from cycle to cycle, as well as index to index. So, we conclude that hysteresis is due to the latitude and meteorological influences as well as solar wind conditions.

The hysteresis magnitude varies non-systematically with the solar cycles, so the inclusion of the hysteresis into the long-term ionospheric predictions seems not suitable.

REFERENCES

Bob Hawke: There is not one outstanding leader in the world" (http://www.smh.com.au/nsw/bob-ha wke-there-is-not-one-outstanding-leader-in-the-world-20160708-gq1kgf.html). Sydney MorningHerald. 2016-07-08.

Blackadder, Jesse (2015). "Frozen Voices: Women, Silence and Antarctica" (http://press-files.anu.edu.au/downloads/press/p316261/pdf/Frozen-voices-Women-silence-and-Antarctica.pdf) (PDF). In Hince, Bernadette; Summerson, Rupert; Wiesel, Arnan (eds.). Antarctica: Music, Sounds, and Cultural Connections. Canberra: ANU Press. p. 90.

Hulbe, Christina L.; Wang, Weili; Ommanney, Simon (2010). "Women in Glaciology, a Historical Perspective" (https://www.igsoc.org/awards/honorary/j10j211.pdf) (PDF). Journal of Glaciology. 56 (200): 947. Retrieved 27 August 2016.

"Women in Antarctica: Sharing this Life Changing Experience" (http://www.development.tas.gov.au/ data/assets/pdf_file/0013/2092/Dr_Robin_Burns_Lecture_-_No._4.pdf), transcript of speech by Robin Burns, given at the 4th Annual Phillip Law Lecture; Hobart, Tasmania, Australia; 18 June 2005. Retrieved 5 August 2010.

"Antarctic Firsts" (http://www.antarctic-circle.org/firsts.htm). Antarctic Circle. 4 October 2014. Retrieved 24 August 2016.

Bogle, Ariel 2016). "New Wikipedia Project Champions Women Scientists in the Antarctic" (http://mashable.com/2016/08/11/wikipedia-antarctic womenscientists/#uLRJOSaqRqqf).Mashable. Retrieved 24 August 2016.

Davis, Amanda (14 April 2016). "This IEEE Fellow Blazed a Trail for Female Scientists in Antarctica" (http://theinstitute.ieee.org/tech-history/technology-history/this-ieee-fellow-blazed-a-trail-for-female-scientists-in-antarctica). The Institute. Retrieved 27 August 2016.

Barr, William (2014). "Review of OF MAPS AND MEN: THE MYSTERIOUS DISCOVERY OF ANTARCTICA". Arctic. 67 (3): 410–411. doi:10.14430/arctic4411 https://doi.org/10.14430%2Farctic4411). JSTOR 24363785

Barros Arana, Diego. "Capítulo XI". Historia general de Chile (http://www.cervantesvirtual.com/obravisor/historia-general-de-chile-tomo-cuarto--0/html/ff2f1efc-82b1-11df-acc7-002185ce6064_67.html)(in Spanish). Tomo cuarto (Digital edition based on the second edition of 2000 ed.). Alicante:Biblioteca Virtual Miguel de Cervantes. p. 280.

Lane, Kris E. (1998). Pillaging the Empire: Piracy in the Americas 1500–1750 (https://books.google.com/?id=bRgFqADzOLkC &printsec=frontcover#v=onepage&q=Brouwer&f=false). Armonk, N.Y.: M.E. Sharpe. p. 88. ISBN 978-0-76560-256-5.

Kock, Robbert. "Dutch in Chile" https://web.archive.org/web/20160229232448/http://www.colonialv oyage.com/dutch-chile/). Colonial Voyage.com. Archived from the original (http://www.colonialvoyag

e.com/dutchchile.html) on 29 February 2016. Retrieved 23 October 2014.

Dalrymple, Alexander. (1771). A Collection of Voyages Made to the Ocean Between Cape Horn and Cape of Good Hope. Two volumes. London.

Headland, Robert K. (1984). The Island of South Georgia, Cambridge University Press. ISBN 0-521-25274-1

Cook, James. (1777). A Voyage Towards the South Pole, and Round the World. Performed in His Majesty's Ships the Resolution and Adventure, In the Years 1772, 1773, 1774, and 1775. In which is included, Captain Furneaux's Narrative of his Proceedings in the Adventure during the Separation of the Ships (http://www.gutenberg.org/

files/15869/15869-8.txt). Volume II. London: Printed for W.Strahan and T. Cadell. (Relevant fragment)

Erki Tammiksaar (14 December 2013). "Punane Bellingshausen" [Red Bellingshausen]. Postimees. Arvamus. Kultuur (in Estonian).

Alan Gurney, Below the Convergence: Voyages Toward Antarctica, 1699–1839, Penguin Books, New York, 1998. p. 181

Bourke, Jane (2004). "Amazing Antarctica: Resource book" (https:// books.google.com/books?id=-Na7Qdz0C&lpg=PA6&dq=Captain %2520John%2520Davis%2520antarctica&pg=PA6#v=onepage &q=Captain%2520John%2520Davis%2520antarctica&f=false). Ready-Ed Publications. Retrieved 2019-06-05.

Charles Wilkes" (http://www.south-pole.com/p0000079.htm). South-Pole.com.

Thomson (1977). "An Annotated Bibliography Of The Paleontology Of Lesser Antarctica And The Scotia Ridge". N.Z. Journal of Geology and Geophysics. 20 (5): 865–904. doi:10.1080/00288306.1977.10420686 (https://doi.org/10.1080%2F00288306.1977.10420686).

"Hero: A New Antarctic Research Ship" (http://www.palmerstation. com/hero/newship.html). PalmerStation.com. 1968.

ANTARCTICEXPLORATION—CHRONOLOGY" https://web.archive. org/web/20060908120017/http://www.quarkexpeditions.com/ antarctica/exploration.shtml). Quark Expeditions. 2004. Archived from the original (http://www.quarkexpeditions.com/antarctica/ exploration.shtml) on 2006-09-08. Retrieved 2006-10-20.

Antarctic Circle—Antarctic First (http://www.antarctic-circle. org/firsts.htm). Antarctic-circle.org. Retrieved on 2012-01-29. https://en.wikipedia.org/wiki/History_of_Antarctica

BAINES, P.G. and CONDIE, S. 1998. Observations and modelling of Antarctic downslope flows: a review. In *Ocean, Ice and Atmosphere: Interactions at the Antarctic Continental Margin*, AGU *Antarctic Research Series* Vol. **75**,

S.S. Jacobs and R. Weiss editors, 29-49.

BRYDEN, H.L. 1979. Poleward heat flux and conversion of available potential energy in Drake Passage, Journal of Marine Research, 37, 1-22.

DEACON, G.E.R. 1937. The hydrology of the Southern Ocean, *Discovery Reports,* **15,** 1-124.

FAHRBACH, E., HOPPEMA, M., ROHARDT, G., SCHRÖDER, M. and WISOTZKI, A. 2004. Decadal-scale variations of water mass properties in the deep Weddell Sea, Ocean Dynamics, 54, 77-91.

HOGG, A.M., MEREDITH, M.P., BLUNDELL, J.R. AND WILSON, C. 2008. Eddy heat flux in the Southern Ocean: Response to variable wind forcing, *Journal of Climate*, **21**(4), 608-620.

HOGG, N.G, 2001. Quantification of the deep Circulation. In: Eds. G. Siedler, J. Church and J. Gould, *Ocean circulation and climate; observing and modelling the global ocean, International Geophysics Series,* **77**, 259-270, Academic Press.

HUGHES, C.W. and ASH, E.R. 2001. Eddy forcing of the mean flow in the Southern Ocean, *Journal of Geophysical Research*, **106**, 2713-2722.

HUGHES, C.W. and ASH, E.R. 2001. Eddy forcing of the mean flow in the Southern Ocean, *Journal of Geophysical Research*, **106**, 2713-2722.

HEYWOOD, K. J., NAVEIRA-GARABATO, A.C., STEVENS, D.P., ANDMUENCH, R.D. 2004. On the fate of the Antarctic Slope Front and the origin of the Weddell Front, *Journal of Geophysical Research,* **109** (C06021), 1-13.

KLATT, O., FAHRBACH, E., HOPPEMA, M. and ROHARDT, G. 2005. The transport of the Weddell Gyre across the Prime Meridian, *Deep-Sea Research II,* **52**, 513-528.

LUMPKIN, R AND SPEER, K. 2007. Global ocean meridional overturning, *Journal of Physical Oceanography,* **37**, 2550-2562.

MCCARTNEY, M.S. and DONOHUE, K.A. 2007. A deep cyclonic gyre in the Australian-Antarctic Basin, *Progress in Oceanography*, **75**, 675-750.

MCCARTNEY, M.S. 1977. Subantarctic Mode Waters. In: Angel, M. (ed) *A voyage of discovery*, Pergamon, New York, 103-119.

MCCARTNEY, M.S. 1982. The subtropical circulation of Mode Waters, *Journal of Marine Research*, **40** (suppl), 427-464.

ORSI, A.H., JOHNSON, G.C. and BULLISTER, J.B. 1999. Circulation, mixing and production of Antarctic Bottom Water, *Prog. Oceanog.*, **43**, 55-109.

ORSI, A.H., WHITWORTH III, T.W. and NOWLIN JR., W.D. 1995. On the meridional extent and fronts of the Antarctic Circumpolar Current, *Deep-Sea Res.*, **42**, 641-673.

RINTOUL, S.R. and SOKOLOV, S. 2001. Baroclinic transport variability of the Antarctic Circumpolar Current south of Australia (WOCE repeat section SR3), Journal of Geophysical Research, 106, 2795-2814.

SCHMITZ, W. J. 1995. On the Interbasin-scale Thermohaline Circulation, *Reviews of Geophysics,* **33**, 2, American Geophysical Union, 151-173.

TRENBERTH, K.E. AND CARON, J.M. 2001. Estimates of meridional atmosphere and ocean heat transports, Journal of Climate, 13, 4358-4365.

TRENBERTH, K.E. ET AL. 2007. Observations: Surface and Atmospheric Change. Contribution of Working Group I to the Fourth Assessment Report of the Intergovernmental Panel on Climate Change,. In: D.Q. S.

Solomon, M. Manning, Z. Chen, M. Marquis, K.B. Averyt, M. Tignor and H.L. Miller (Editor), Climate Change 2007: The Physical Science Basis. Cambridge University Press,, Cambridge, U.K., and New York, U.S.A.

SOKOLOV, S. and RINTOUL, S.R. 2002. The structure of Southern Ocean fronts at 140E, Journal of Marine Systems, 37, 151-184.

SABINE, C.L., ET AL. 2004. The Oceanic Sink for Anthropogenic CO_2, *Science*, **305**, 367-371.

WÜST, G. 1935. The Stratosphere of the Atlantic Ocean. *Scientific Results of the German Atlantic Expedition of the Research Vessel "Meteor" 1925 – 1927* (English translation, E.J. Emery (ed.), Amerind, new Delhi, 1978), **6**, 109-288.

KING, J.C. and TURNER, J. 1997. Antarctic meteorology and climatology, Cambridge University Press, Cambridge, UK, 409 pp.

KING, J.C. 1994. Recent variability in the Antarctic Peninsula, *International Journal of Climatology,* **14**(4), 357-369.

JACKA, T.H. and BUDD, W.F. 1991. Detection of temperature and sea ice extent changes in the Antarctic and Southern Ocean.

Weller, G., Wilson, C. L., and Severin, B. A. *Proceedings of the International Conference on the Role of the Polar Regions in Global Change. June 11-15, 1990, University of Alaska Fairbanks,* 63-70. Fairbanks, AK, University of Alaska, Geophysical Institute.

JACKA, T.H. and BUDD, W.F. 1998. Detection of temperature and sea-ice-extent changes in the Antarctic and Southern Ocean, 1949-96, *Annals of Glaciology,* **27**, 553-559.

JONES, P.D. 1995. Recent variations in mean temperature and the diurnal temperature range in the Antarctic, *Geophysics Research Letters,* **22** [11], 1345- 1348.

MARSHALL, G. J. 2003. Trends in the Southern Annular Mode from Observations and Reanalyses, *Journal of Climate*, **16**, 4134-4143.

MARSHALL, G. J., ORR, A., VAN LIPZIG, N.P.M. and KING, J.C. 2006. The impact of a changing Southern Hemisphere Annular Mode on Antarctic Peninsula summer temperatures, *Journal of Climate*, **19** [20], 5388-5404.

MEEHL, G.A., T.F. STOCKER, W.D. COLLINS, P. FRIEDLINGSTEIN, A.T. GAYE, J.M. GREGORY, A. KITOH, R. KNUTTI, J.M. MURPHY, A. NODA, S.C.B. RAPER, I.G. WATTERSON, A.J. WEAVER, and Z.-C. ZHAO, 2007:

Global Climate Projections. In: *Climate Change 2007: The Physical Science Basis. Contribution of Working Group I to the Fourth Assessment Report of the Intergovernmental Panel on Climate Change.* Cambridge University Press, Cambridge, United Kingdom and New York, NY, USA, 749-845

RAPER, S.C., WIGLEY, T.M., JONES, P.D. and SALINGER, M. J. 1984. Variations in surface air temperatures: Part 3. The Antarctic, 1957-1982, *Monthly Weather Review,* **112**, 1341-1353.

RAPHAEL, M.N. and HOLLAND, M.M. 2006. Twentieth century simulation of the Southern Hemisphere climate in coupled models. Part 1: Large scale circulation variability, *Clim. Dyn.,* **26**, 217–228, doi:10.1007/s00382-005-0082-8.

RAPHAEL, M.N. 2003. Impact of observed sea-ice concentration on the Southern Hemisphere extratropical atmospheric circulation in summer, *J. Geophys. Res.,* **108**, No. D22, 4687 10.1029/2002JD003308.

SOKOLOV, S. and RINTOUL, S.R. 2003. The subsurface structure of interannual temperature anomalies in the Australian sector of the Southern Ocean, *Journal of Geophysical Research,* **108** (C9): Art. No. 3285.

FOGT R.L., PERLWITZ, J., PAWSON, S., and OLSEN, M.A. 2009a. Intra-annual relationships between polar ozone and the SAM, *Geophys. Res. Lett.,* **36**, L04707, doi:10.1029/2008GL036627.

FOGT, R.L. and BROMWICH, D.H. 2006. Decadal Variability of the ENSO Teleconnection to the High-Latitude South Pacific Governed by Coupling with the Southern Annular Mode, *J. Climate,* **19**, 979–997.

FOGT, R.L., PERLWITZ, J., MONAGHAN, A.J., BROMWICH, D.H., JONES, J.M. and MARSHALL, G.J. 2009b. Historical SAM Variability. Part II: 20th century variability and trends from reconstructions, observations, and the IPCC AR4 models, *J. Climate,* In Press.

STEIG, E.J., SCHNEIDER, D.P., RUTHERFORD, S.D., MANN, M.E., COMISO, J.C. and SHINDELL, D.T. 2009. Warming of the Antarctic ice-sheet surface since the 1957 International Geophysical Year, *Nature*, **457**, 459-462.

BROMWICH, D.H., FOGT, R.L., HODGES, K.I. and WALSH, J.E. 2007. A tropospheric assessment of the ERA-40, NCEP, and JRA-25 global reanalyses in the polar regions, *J. Geophys. Res.*, **112,** Art. No. D10111.

MARSHALL, A., LYNCH, A.H. and GÖRGEN, K. 2007. A model intercomparison study of the Australian summer monsoon. Part II: Variability at intraseasonal and interannual timescales, *Aust. Met. Mag.* (in review).

BROMWICH, D.H., FOGT, R.L., HODGES, K.I. and WALSH, J.E. 2007. A tropospheric assessment of the ERA-40, NCEP, and JRA-25 global reanalyses in the polar regions, *J. Geophys. Res.*, **112,** Art. No. D10111.

BRACEGIRDLE, T.J., CONNOLLEY, W.M. and TURNER, J. 2008. Antarctic climate change over the Twenty First Century, *Journal of Geophysical Research Atmospheres,* **113**, D03103, doi:03110.01029/02007JD008933.

BRACHFELD, S., DOMACK, E., KISSEL, C., LAJ, C., LEVENTER, A., ISHMAN, S., GILBERT, R., CAMERLENGHI, A. and EGLINTON, L.B. 2003. Holocene history of the Larsen-A Ice Shelf constrained by geomagnetic paleointensity dating, *Geology*, **31**, 749-752.

JONES, A.E., ANDERSON, P.S., WOLFF, E.W., TURNER, J., RANKIN, A.M. and COLWELL, S.R. 2006. A role for newly-forming sea ice in springtime polar tropospheric ozone loss? Observational evidence from Halley station, Antarctica, *J. Geophys. Res.*, **111**, D08306, doi:10.1029/2005JD006566

TURNER, J. 2004. The El Niño-Southern Oscillation and Antarctica, *International Journal of climatology*, **24**, 1-31.

TURNER, J., LACHLAN-COPE, T.A., COLWELL, S.R., MARSHALL, G.J. and CONNOLLEY, W.M. 2006. Significant warming of the Antarctic winter troposphere, *Science*, **311,** 1914-1917.

TURNER, J., COLWELL, S. R., MARSHALL, G. J., LACHLAN-COPE, T. A., CARLETON, A. M., JONES, P. D., LAGUN, V., REID, P. A. and

IAGOVKINA, S. 2004. The SCAR READER project: Towards a high-quality database of mean Antarctic meteorological observations, *Journal of Climate*, **17**, 2890-2898.

TURNER, J., COLWELL, S.R., MARSHALL, G.J., LACHLAN-COPE, T.A., CARLETON, A.M., JONES, P.D., LAGUN, V., REID, P.A. and IAGOVKINA, S. 2005a. Antarctic climate change during the last 50 years, *International Journal of Climatology*, **25**, 279-294.

TURNER, J., COMISO, J.C., MARSHALL, G.J., LACHLAN-COPE, T.A., BRACEGIRDLE, T., MAKSYM, T., MEREDITH, M.P., WANG, Z. and ORR, A. 2009. Non-annular atmospheric circulation change induced by stratospheric ozone depletion and its role in the recent increase of Antarctic sea ice extent, *Geophys. Res. Lett.*, **36**, L08502, doi:10.1029/2009GL037524.

CHAPMAN.W.L. and WALSH, J. E. 2007. A Synthesis of Antarctic Temperatures, *Journal of Climate*, **20**, 4096-4117.

Sakaguchi, K., K. Shiokawa, A. Ieda, Y. Miyoshi, Y. Otsuka, and T. Ogawa "Simultaneous ground and satellite observations of an isolated proton arc at sub-auroral latitudes". Journal of Geophysical Research. 2007. Retrieved 5 August 2015.

Siscoe, G. L. (1986). "An historical footnote on the origin of 'aurora borealis'". History of Geophysics: Volume 2. History of Geophysics: Volume 2. Series: History of Geophysics. History of Geophysics. 2. pp. 11–14. Bibcode:1986HGeo....2...11S. doi:10.1029/HG002p0011. ISBN 978-0-87590-276-0.

Østgaard, N.; Mende, S. B.; Frey, H. U.; Sigwarth, J. B.; Åsnes, A.; Weygand, J. M. (2007). "Auroral conjugacy studies based on global imaging". Journal of Atmospheric and Solar-Terrestrial Physics. 69 (3): 249. Bibcode:2007JASTP..69..249O. doi:10.1016/j.jastp.2006.05.026

Feldstein, Y. I. (2011). "A Quarter Century with the Auroral Oval". EOS. 67 (40): 761. Bibcode:1986EOSTr..67..761F. doi:10.1029/ EO067i040p00761-02.

Stamper, J.; Lockwood, M.; Wild, M. N. (1999). "Solar causes of the long-term increase in geomagnetic activity" (PDF). J. Geophys. Res. 104 (A12): 28, 325–28, 342. Bibcode:1999JGR...10428325S. doi:10.1029/1999JA900311.

Papitashvili, V. O.; Papitashva, N. E.; King, J. H. (2000). "Solar cycle effects in planetary geomagnetic activity: Analysis of 36-year long OMNI dataset" (PDF). Geophys. Res. Lett. 27 (17)2797–2800. Bibcode:2000GeoRL..27.2797P.doi:10.1029/2000GL000064 hdl:2027.42/94796

Bryant, Duncan (1998). Electron-Acceleration-in-the-Aurora-and-Beyond. Bristol & Philadelphia: Institute of Physics Publishing Ltd. p. 163. ISBN 978-0750305334.

Lev Davidovich Landau". history.mcs.st-andrews.ac.uk.Alfvén, Hannes (3 October 1942). "Existence of electromagnetic-hydrodynamic waves". Nature. 150 (3805): 405–406. Bibcode:1942Natur.150..405A. doi:10.1038/150405d0

Aarons J., (1982). Global morphology of ionospheric scintillations. Proc. IEEE 70 (4), 360-378.

Aarons J., Lin B., Mendillo M. and Liou K., (2000). Codrescu M., Global Positioning System phase fluctuations and ultraviolet images from the Polar satellite. Journal of Geophysical Research, Vol. 105, A3, pp.5201-5213.

Aarons J., Mendillo M. and Yantosca R., (1997). GPS phase fluctuations in the equatorial region during sunspot minimum. Radio Science, Vol. 32, No 4, pp. 1535-1550.

Aarons J., Mullen J. P. and Whitney, (1981). H. E. UHF scintillation activity over polar latitudes. Geophysics. Res. Lett. 8 (3), 277-280.

Basler R. P. and DeWitt R. N., (1962). The height of ionospheric irregularities in the auroral zone. J. Geophys. Res. 67, 587-593.

Basu S., Basu S. and Aarons J., (1978). McClure J, Cousins M. On the coexistence of kilometer- and meter-scale irregularities in the Nighttime Equatorial F region. J Geophys Res 83(A9):4219-4226.

Basu S., Basu S., Senior C., Weimer D., Nielsen E. and Fougere P., (1986). Velocity shears and sub-km scale irregularities in the Nighttime Auroral F-region. Geophys Res Lett 13(2):101-104.

Basu S., Bssu S., Weber E. J. and Coley W. R., (1988). Case study of polar cap scintillation modeling using DE-2 irregularity measurements at 800 km. Radio Sci. 23, 545.

Basu S., MacKenzie E., Basu S., Carlson H., Hardy D., Rich F. and Livingston R., (1983). Coordinated measurements of low-energy electron precipitation and scintillations/TEC in the auroral oval. Radio Sci 18(6):1151-1165.

Basu S. and Valladares C., (1999). Global aspects of plasma structures. J. Atmos. Solar Terr, Phys. 61, 127-139.

Basu S., Weber E., Bullett T., Keskinen M., MacKenzie E., Doherty P., Sheehan R., Kuenzler H., Ning P. and Bongiolatti J., (1998). Characteristics of plasma structuring in the cusp/cleft region at Svalbard. Radio Sci 33(6):1885-1899.

Coker C., Hunsucker R. and Lott G., (1995). Detection of auroral activity using GPS satellites, Geophysical Research Letters, Vol. 22, No 23, pp. 3259-3262.

Da Rasa A. V., Waldman H., Bendito J. and Garriott O. K., (1973). Response of the ionospheric electron content to fluctuations in solar activity. J. Atmos. Terr. Phys. 35, 1429-1442.

De Franceschi G., Alfonsi L. and Romano V., (2006). Isacco an Italian project to monitor the high latitude ionosphere by means of GPS receivers. GPS Solut 10:263-267. doi:10.1007/s10291-006- 0036-6.

De Franceschi G., Alfonsi L., Romano V., Aquino M., Dodson A., Mitchell C. N., Spencer P. and Wernik A. W., (2008). Dynamics of high latitude patches and associated small-scale irregularities during the October and November 2003 storms. J Atmos Solar-Terr Phys 70:879-888.

Doherty P., Raffi E., Klobuchar J. and El-Arini M. B., (1994). Statistics of time rate of change of ionospheric range delay, In: Proceedings of ION GPS-94, Part 2, Salt Lake City, 1589 pp, 1994.

Fejer B. G., Scherlies L. and de Paula E. R., (1999). Effects of the vertical plasma drift velocity on the generation and evolution of equatorial spread F. J Geophys Res 104:19859-19869.

Feitcher E. and Leitinger R. A., (1997). 22-year cycle in the F-layer ionization of the ionosphere. Ann. Geophysicae 15, 1015-1027.

Forte B. and Radicella S. M., (2002). A different approach to the analysis of GPS scintillation data. Ann. Geophysicae 45(3-4), 551-561.

Gendt G. and Dick G., (1995). Workshop Proceedings, Special Topics and New Direction. edited, pp. 57-66, Potsdam.

Jayachandran P. T., Langley R. B., MacDougall J. W., Mushini S. C., Pokhotelov D., Hamza A. M., Mann I. R., Milling D. K., Kale Z. C., Chadwick R., Kelly T., Danskin D. W. and Carrano C. S., (2009). Canadian High Arctic Ionospheric Network (CHAIN). Radio Sci., 44, RS0A03, doi:10.1029/2008RS004046.

Krankowski A., Baran L. W. and Shagimuratov I. I., (2002). Influence of the Northern Ionosphere on Positioning Precision, Physics and Chemistry of the Earth. Vol. 27, pp. 391-395.

Krankowski A., Shagimuratov I. I., Baran L. W. and Ephishov I. I., (2005). Study of TEC fluctuations in Antarctic ionosphere during storm using GPS observations. Acta Geophysica Polonica, Vol. 53, No 2, pp. 205-218.

Krankowski A. and Shagimuratov I., (2006) Impact of TEC Fluctuations in the Antarctic ionosphere on GPS positioning oczapowski St. 1,10-957. Russia Artificial Satellites 41(1), Doi: 10.2478/V10018-007-0005-5.

Kersley L., Pryse S. and Wheadon N., (1988). Amplitude and phase scintillation at high latitudes over northern Europe. Radio Sci 23(3):320-330.

Kersley L., Russell C. D. and Rice D. L., (1995). Phase Scintillations and irregularities in the northern polar ionosphere. Radio Science 30, pp. 619-629.

Kelley M. C., (1989). The earth's ionosphere, plasma physics and electrodynamics. Academic, San Diego.

Lansinger J. M. and Fremeuw E. J., (1967). The scale size of scintillation-producing irregularities in the auroral ionosphere. J. Atmos. Terr. Phys. 29, 1229-1242.

Liu L., Wan W., Ning B. and Zhang M. L., (2009). Climatology of the mean total electron content derived from GPS global ionospheric maps. J. Geophys. Res., 114, A06308, doi: 10.1029/2009JA014244.

Liu L., Zhao B., Wan W., Ning B., Zhang M. L. and He M., (2009). Seasonal variations of the ionospheric electron densities retrieved from Constellation Observing System for Meteorology, Ionosphere, and Climate mission radio occultation measurements. J. Geophys. Res., 114, A02302, doi: 10.1029/2008JA013819.

Meggs R. W., Mitchell C. N. and Honary F., (2008). GPS scintillation over the European Arctic during the November 2004 storms. GPS Solut 12:281-287. doi:10.1007/s10291-008-0090-3.

Mitchell C. N., Alfonsi L., De Franceschi G., Lester M., Romano V. and Wernik A. W., (2005). GPS TEC and scintillation measurements from the polar ionosphere during the October 2003 storm. Geophys Res Lett 32:L12S03. doi:10.1029/2004GL021644.

Natali M. P. and Meza A., (2010). Annual and semiannual VTEC effects at low solar activity based on GPS observations at different geomagnetic latitudes. J. Geophys. Res., 115, D18106, doi: 10.1029/2010JD014267.

Pi X., Manucci A. J., Lindqwister U. J. and Ho C. M., (1997). Monitoring of global ionospheric irregularities using the worldwide GPS network, Geophysical Research Letters Vol 24, No18, pp.2283-2286.

Rino C. L. and Matthews S. J., (1980). On the morphology of auroral zone radio wave scintillation. J. Geophys. Res. 85, p. 4139.

Rodger A. S., Pinnock M., Dudney J. R., Baker K. B. and Greenwald R. A., (1988). A new mechanism for polar patch formation. J. Geophys. Res. 99, 6425-6436.

Soicher H., (1988). Traveling ionospheric disturbances (TIDs) at mid-latitude: Solar cycle phase dependence. Radio Sci. 23, 283-291.

Sojka J. J., Bowline M. D., Schunk R. W., Decker D. T., Valladares C. E., Sheehan R., Anderson D. N. and Heelis R. A., (1993). Modeling polar cap F-region patches using time varying convection. Geophys. Res. Lett. 20(17), 1783- 1786.

Tsunoda R. T., (1988). High-latitude F-region irregularities: A review and synthesis. Rev. Geophys. 26, 719-760.

Valladares C. E., Basu S., Buchau J. and Friis-Christiansen E., (1994). Experimental evidence for the formation and entry of patches into the polar cap. Radio Sci. 29, 167-194.

Van Velthoven P. J., (1990). Medium-scale irregularities in the ionospheric electron content. Ph.D. Thesis, Technische Universiteit Eindhoven.

Wanninger L., (1993). The occurrance of ionospheric disturbances above Japan and their effects on GPS positioning. Proceedings of the 8th International Symposium on Recent Crust Movemens (CRCM 93), pp.175-179, Kobe, Japan, December 6-11.

Wanninger L., (1995). Monitoring ionospheric disturbances using IGS Network, Proc. Of the 1995 IGS workshop, Potsdam, Germany.

Weber E. J., Buchau J., Moore J. G., Sharber J. R., Livingston R. C., Winningham J. D., Reinisch B. W., (1984). F-layer ionization patches in the polar cap. J. Geophys. Res. 89, 1683-1694.

Weber E. J., Klobuchar J. A., Buchau J., Carlson H. C., Livingston R. C., Beaujardiere O. de la, McCreday M., Moore J. G. and Bishop J. G., (1986). Polar cap F-layer patches: Structure and dynamics. J. Geophys, Res. 91, 2121-2129.

Whitney H. E., Aarons J. and Malik C., (1969). A proposed index for measuring ionospheric scintillations. Planet. Space Sci. 17, 1069-1073.

Aarons J., (1982). Global morphology of ionospheric scintillations. Proc. IEEE, 70 (4), 360-378.

Aarons J., (1982). Global morphology of ionospheric scintillation. Proc. IEEE 70, 360-378.

Aarons J., (1993). The longitudinal morphology of equatorial F-layer irregularities relevant to their occurrence. Space Sci. Rev. 63, 209-243.

Aarons J., Mendillo M., Yantosca R. and Kudeki E., (1996). GPS Phase Fluctuations in the Equatorial Region during the MISETA Campaign. Journal of Geophysical Research, 101, pp. 26851-26862.

Aarons J., Mullen J. P. and Whitney H. E., (1969). The scintillation boundary. J. Geophys. Res., 74, 884-889.

Aarons J., Mullen J. P. and Whitney H. E., (1981). UHF scintillation activity over polar latitudes. Geophysics. Res. Lett., 8 (3), 277-280.

Alfonsi L., Materassi M. and Wernik A.W., (2003). Distribution of scintillation parameters calculated from in situ data – preliminary results. Proceedings of "Atmospheric Re-mote Sensing using Satellite Navigation Systems", Special Symposium of the URSI Joint Working Group FG, Matera.

Basler R. P. and DeWitt R. N., (1962). The height of ionospheric irregularities in the auroral zone. J. Geophys. Res., 67, 587-593.

Basu S. and Basu S., (1981). Equatorial scintillations a review. J. Atmos. Terr. Phys., 43 (5), 473-489.

Basu S. and Basu S., (1985). Equatorial scintillations: Advances since ISEA-6. J. Atmos. Terr. Phys. 47, 753-768.

Basu S. and Basu S., (1993). Ionospheric structures and scintillation spectra. In: V.I. Tatarski, A. Ishimaru and V.U. Zavorotny (eds.), "Wave Propagation in Random Media (Scintillation)", pp. 139-153, The International Society for Optical Engineering, Belling-ham, WA, USA.

Basu S., Basu S., Livingston R. C., Whitney H. E. and Mac-Kenzie E., (1981). Comparison of ionospheric scintillation statistics from the North Atlantic and Alaskan sectors of the auroral oval using the WIDEBAND satellite. Air Force Geophys. Lab., Hanscom AFB, MA, Rep. AFGL-TR-81-0266, AD A111871, September.

Basu S., Basu S., McClure J. P., Hanson W. B. and Whitney H. E., (1983). High-resolution topside in-situ data of electron densities and VHF/GHz scintillations in the equatorial region. J. Geophys. Res., 88, 403-415.

Basu S., Groves K. M., Basu Su. and Sultan P. J., (2002). Specification and forecasting of scintillations in communication/navigation links: current status and future plans. J. Atmos. Solar-Terr. Phys. 64, 1745-1754.

Booker H. G., Ratcliffe J. A. and Shinn D. H., (1950). Diffraction from an irregular screen with application to ionospheric problems. Phil. Trans. Roy. Soc. A. 242, 579.

Briggs B. H. and Parkin I. A., (1963). On the variation of radio star and satellite scintillation with zenith angle. J. Armos. Terr. Phys., vol. 25, pp. 339-365.

Fejer B. G., (1996). Natural ionospheric plasma waves. In: H. Kohl, R. Ruster and K. Schlegel (eds.), "Modern Ionospheric Science", pp. 216-273, European Geophysical Society, Katlenburg-Lindau, FRG.

Franke S. J. and Liu C. H., (1981). Observations and modeling of multi- frequency VHF and GHz scintillations in the equatorial region. J. Geophys. Res., vol. 88, p. 7075, 1983. VOI. 16, pp.939-945.

Fremouw E. J., Livingston R. C. and Miller D. A., (1980). On statistics of scintillating signals. J. Atmos. Terr. Phys. 42, 717-731.

Fremouw E. J. and Secan J. A., (1984). Modelling and scientific application of scintillation results. Radio Sci. 19, 687-694.

Groves K. M., Basu S., Weber E. J., Smitham M., Kuenzler H., Valladares C. E., Sheehan R., Mackenzie E., Secan J. A., Ning P., McNeill W. J., Moonan D. W. and Kendra M. J., (1997). Equatorial scintillation and systems support. Radio Sci. 32, 2047-2064.

Heppner J. P., Liebrecht M.C., Maynard N.C. and Pfaff R. F., (1993). High-latitude distributions of plasma waves and spatial irregularities from DE 2 alternating current electric field observations. J. Geophys. Res. 98, 1629-1652.

Hewish A., (1951). The diffraction of radio waves in passing through a phase-changing iono-sphere. Proc. Roy. Soc. A. 209, 81.

Huba J. D., (1989). Theoretical and simulation methods applied to high latitude, F region turbu-lence. In: C.H. Liu (ed.), "World Ionosphere/ Thermosphere Study", WITS Handbook, vol. 2, pp. 399-428, SCOSTEP Secretariat, Boulder, CO.

Kelley M. C., (1989). The Earth Ionosphere. Academic Press, London.

Kersley L., Russell C. D. and Rice D. L., (1995). Phase Scintillations and Irregularities in the Northem Polar Ionosphere. Radio Science, 30, pp. 619-629.

Keskinen M. J. and Ossakow S. L., (1983). Theories of high-latitude irregularities: A review. Radio Sci. 18, 1077-1092.

Kintner P. M. and Seyler C. E., (1985). The status of observations and theory of high latitude ionospheric and magnetospheric plasma turbulence. Space Sci. Rev. 41, 91-129.

Knepp D. L., (1983). Multiple phase-screen calculation of the temporal behavior of stochastic waves. Proc. IEEE 71, 722-737.

Lansinger J. M. and Fremeuw E. J., (1967). The scale size of scintillation-producing irregularities in the auroral ionosphere. J. Atmos. Terr. Phys., 29, 1229-1242.

Ratcliffe J. A., (1956). Some aspects of diffraction theory and their application to the iono-sphere. Rep. Progr. Phys. 19, 188-267.

Rino C. L., (1979). A power law phase screen model for ionospheric scintillation, 2, Strong scatter. Radio Sci. 14, 1147-1155.

Rino C. L., (1980). Numerical computations for a one-dimensional power law phase screen. Radio Sci. 15, 41-47.

Rino C. L., (1982). On the application of phase screen models to the interpretation of iono-spheric scintillation data. Radio Sci. 17, 855-867.

Rino C. L. and Fremouw E. J., (1977). The angle dependence of singly scattered wavefields. J. Atmos. Terr. Phys. 39, 859-868.

Rino C. L. and Matthews S. J., (1980). On the morphology of auroral zone radio wave scintillation. J. Geophys. Res., vol. 85, p. 4139.

Secan J. R., Bussey R. M. and Fremouw E.J., (1997). High-latitude upgrade to the Wideband ionospheric scintillation model. Radio Sci. 32, 1567-1574.

Secan J. R., Bussey R. M., Fremouw E.J. and Basu S., (1995). An improved model of equatorial scintillation. Radio Sci. 30, 607-617.

Smerd S. F. and Slee O. B., (1966). Regular Variations in the Scintillations of Radio Sources with Season, Time of Day and Solar Distance. Australian Journal of Physics, 19, pp. 427-439.

Tatarski V. I., (1971). The effects of the turbulent atmosphere on wave propagation. Natl. Tech. Inform. Serv., Springfield, VA.

Tsunoda R. T., (1988). High latitude F region irregularities: A review and synthesis. Rev. Geo-phys. Space Phys. 26, 719-760.

Wernik A. W., (1997). Wavelet transform of nonstationary ionospheric scintillation. Acta Geo-phys. Pol. 45, 237-253.

Wernik A.W., Liu C. H. and Yeh K. C., (1980). Model computation of radio wave scintillation caused by equatorial ionospheric bubbles. Radio Sci. 15, 559-572.

Wernik A. W., Secan J. A. and Fremouw E.J., (2003). Ionospheric irregularities and scintilla-tion. Adv. Space Res. 31, 4, 971-981.

Whitney H. E., Aarons J. and Malik C., (1969). A proposed index for measuring ionospheric scintillations. Planet. Space Sci., 17, 1069-1073.

Yeh K. C., and Wernik A.W., (1993). On ionospheric scintillation. In: V.I. Tatarski, A. Ishimaru and V.U. Zavorotny (eds.), "Wave Propagation in Random Media (Scintillation)", pp. 34-49, The International Society for Optical Engineering, Bellingham, WA, USA.

Yeh K. C. and Liu C. H., (1979). Ionospheric effects on radio communication and ranging pulses. IEEE Trans. Antennas and Prop. AP-27, 747-751.

Yeh K. C. and Liu C.H., (1982). Radio wave scintillations in the ionosphere. Proc. IEEE 70, 324-360.

Araujo-Pradere E. A., Fuller-Rowell T. J. and Codrescu M. V., (2002). STORM: an empirical storm time ionospheric correction model 1. Model description. Radio Science 37.

Atac T. and Ozguc A., (2001). Flare index during the rising phase of solar cycle 23. Solar Physics 198, 399–407.

Atac T. and Ozguc A., (2006). Overview of the solar activity during solar cycle 23. Solar Physics 233, 139–153.

Field P. R. and Rishbeth H., (1997). The response of the ionospheric F2-layer to geomagnetic activity: an analysis of worldwide data. Journal of Atmospheric and Terrestrial Physics 59, 163–180.

Forbes J. M., Palo S. E. and Zhang X., (2000). Variability of the ionosphere. Journal of Atmospheric and Terrestrial Physics, 62, 685–693.

Fuller-Rowell T. J., Codrescu M. C. and Wilkinson P., (2000). Quantitative modeling of the ionospheric response to geomagnetic activity. Annales Geophysicae 18, 766–781.

Hathaway D. H., (1996). Doppler Measurements of the Sun's Meridional Flow. Astrophys. J.460, 1027.

Hathaway D. H., Nandy D., Wilson R. M. and Reichmann E. J. (2003). Evidence That a Deep Meridional Flow Sets the Sunspot Cycle Period Astrophysical Journal, Volume 589, Page 665.

Huang Y. N., (1963). The hysteresis variation of the semi-thickness of the F2-layer and its relevant phenomenon at Kokobunji, Japan. Journal of Atmospheric and Terrestrial Physics 25, 647–658.

Kane R. P., (1992). Sunspots, solar radio noise, solar EUV and ionospheric foF2. Journal of Atmospheric and Terrestrial Physics 54 (3/4), 463–466.

Kane R. P., (2006). Are the double-peaks in solar indices during solar maxima of cycle 23 reflected in ionospheric foF2. Journal of Atmospheric and Terrestrial Physics 68, 877–880.

Kane R. P., (2009). Fluctuations of solar activity during the declining phase of the 11 year sunspot cycle. Solar Physics, 255, 163-168.

Kleczek J., (1952). Solar Flare Index. Publication of Astrophysical Observatory, No. 24, Prague.

Lakshimi D. R., Reddy B. M. and Dabas R. S., (1998). On the possible use of recent EUV data for ionospheric prediction. Journal of Atmospheric and Terrestrial Physics 50 (3), 207–213.

Muggleton L. M., (1969). Secular variation in F-region response to sunspot number. Journal of Atmospheric and Terrestrial Physics 31, 1413–1419.

Naismith R., Bevan H. C. and Smith P. A., (1961). A long term variation in the relationship of sunspot numbers to E-region character figures. Journal of Atmospheric and Terrestrial Physics 21, 167–173.

Naismith R. and Smith P. A., (1961). Further evidence of a long-term variation in the relationship of solar activity to the ionosphere. Journal of Atmospheric and Terrestrial Physics 22, 270–274.

Ortiz de Adler N. and Elias A. G., (2008). Latitudinal variation of foF2 hysteresis of solar cycle 20, 21, and 22 and its application to the analysis of long-term trends. Ann. Geophys. 26, 1269-1273.

Ozguc A., Atac T. and Rybak J., (2002). Flare index variability in the ascending branch of solar cycle 23. Journal of Geophysical Research 107 (A7) SHH11-1-8.

Ozguc A., Atac T. and Rybak J., (2003). Temporal variability of the flare index (1966–2001). Solar Physics 214, 375–397.

Ozguc. A., Atac T. and Pektas R., (2007). Examination of the solar variation of foF2 for cycle 22 and 23. journal of Atmospheric and solar terrestrial physics, vol. 70, 268-276.

Ozguc A., Tulunay Y. and Atac T., (1998). Examination of the solar cycle variation of foF2 by using solar flare index for the cycle 21. Advances in Space Research 22 (1), 139–142.

Rao M. S. V. G. and Rao R. S., (1969). The hysteresis variation in F2-layer parameters. Journal of Atmospheric and Terrestrial Physics 31, 1119–1125.

Rishbeth H. and Field, P. R., (1997). Latitude and solar-cycle patterns in the response of the ionosphere F2-layer to geomagnetic activity. Advances in Space Research 20, 1689–1692.

Sethi N. K., Goel M. K. and Mahajan K. K., (2002). Solar cycle variations of foF2 from IGY to 1990. Ann. Geophysicae, 20, 1677-1685.

Smith P. A. and King J. W., (1981). Long-term relationships between sunspots, solar faculae and the ionosphere, J. Atmos.Terr. Phys. 43, 1057 – 1063.